psychology in crisis

Also by Brian M. Hughes

Conceptual and Historical Issues in Psychology (2012, Prentice-Hall)

Rethinking Psychology: Good Science, Bad Science, Pseudoscience (2016, Palgrave)

Psychology in Crisis

Brian M. Hughes

National University of Ireland, Galway

First published 2018 by
PALGRAVE

Palgrave in the UK is an imprint of Macmillan Publishers Limited, registered in England, company number 785998, of 4 Crinan Street, London, N1 9XW.

Palgrave® and Macmillan® are registered trademarks in the United States, the United Kingdom, Europe and other countries.

ISBN 978–1–352–00300–0 paperback

This book is printed on paper suitable for recycling and made from fully managed and sustained forest sources. Logging, pulping and manufacturing processes are expected to conform to the environmental regulations of the country of origin.

A catalogue record for this book is available from the British Library.

A catalog record for this book is available from the Library of Congress.

Contents

Acknowledgements   vi

1   *'The Same Again, But Different'*: Psychology's Replication Crisis   1

2   *'Black Is White'*: Psychology's Paradigmatic Crisis   28

3   *'Never Mind the Quality, Feel the Width'*: Psychology's Measurement Crisis   46

4   *'That Which Can Be Measured'*: Psychology's Statistical Crisis   73

5   *'We Are The World'*: Psychology's Sampling Crisis   102

6   *'Fitter, Happier, More Productive …'*: Psychology's Exaggeration Crisis   119

7   From Crisis to Confidence: Dealing with Psychology's Self-Inflicted Crises   144

References   173

Index   190

Acknowledgements

Thank you to everyone at Palgrave, past and present, for their work on this book. I pay particular tribute to Luke Block, Stephanie Farano, Cathy Scott, and Paul Stevens.

I am fortunate to have worked with many fine collaborators and friends over the years, who have undoubtedly influenced the thinking that appears on these pages. Among them I thank Douglas Carroll, Ann-Marie Creaven, Stephen Gallagher, Lokesh Joshi, Ad Kaptein, Łukasz Kaczmarek, Krys Kaniasty, Terri Morrissey, Lynn Myers, Donncha O'Connell, Aoife O'Donovan, Páraic Ó Súilleabháin, Albert Sesé, Michael Smith, Mark Wetherell, and Anna Whittaker. More verbose gratitude goes to Siobhán Howard for feedback on statistical points and for curtailing some of my more unwise impulses, to Wei Lü for input on methods, language, literature, habituation, psychophysiology, and emotional regulation, and to Jack James for providing mentorship and role-modelling throughout my academic career.

Thank you to Nóirín Buckley and Paul O'Donoghue for fostering my contrarianism at an early stage and to Arnold Byrne and Carolyn Wilshire for advice and materials.

Thanks also, for their ongoing forbearance and patience, to my family, the various generations of which, in ascending order, include: Annie and Louis; Marguerite; and Jarlath and Mary (and Nora).

B. M. H.

CHAPTER 1

'The Same Again, But Different': Psychology's Replication Crisis

Hurricane in a teacup

Stop the presses! We bring you some breaking news:

Scientists Say Female Hurricanes are DEADLIER than Male Hurricanes!

Could there be a more twenty-first century headline? Its synergy of human-interest soundbites – gender politics, life and death, even the *weather* – might just represent the epitome of clickbait.

And the news conveyed is so seductively counterintuitive. Or is it intuitive? I suppose it depends on your prejudices. So let us contemplate its main point: according to research, if a hurricane has a woman's name, it becomes *more deadly*. It literally *kills more people*. Hurricane Eve is a greater threat than Hurricane Steve.

But it's a *hurricane*, you might say. How could this be true? How could the name human beings choose for it cause a weather system to wreak a different calibre of havoc? The idea seems, well, implausible.

Nonetheless, the smoke of this news, which made global headlines in 2014, came from the fire of science. Professional researchers reported the claim to be true. More specifically, *psychology* researchers reported it. And psychology, lest we forget, is a science. Psychologists use scientific methods. Psychologists are committed to a search for scientific truth. They wouldn't seek to publish such a finding unless it was reliable, would they?

Well – spoiler alert – as we will go on to see throughout this book, when it comes to psychology, it is never quite safe to assume anything.

The hurricanes-are-deadlier-than-himmicanes study originally appeared in the prestigious American academic journal *Proceedings of the National Academy of Sciences* (Jung et al., 2014). Its authors presented data showing that more people are killed by female hurricanes than by male ones. A lot more. According to them:

> changing a severe hurricane's name from Charley ... to Eloise ... could nearly triple its death toll. (p. 8783)

To be fair, they did not claim that the wind blows differently when it has a girly name. They argued that *humans* react differently depending on the gender they perceive a hurricane to be. Humans take female hurricanes less seriously than male ones. It isn't a gender effect so much as it is a gender *bias*. So addled are people by sexist prejudice, they even stereotype the weather. In the words of one CNN columnist (Cupp, 2014):

> *Girl hurricanes get no respect.*

The news spread widely through traditional and social media. And why not? The finding was not just intriguing, it appeared in a highly reputable journal, maybe even one of the most reputable journals in all of science.

But there was, of course, a problem. The finding – while earnestly presented, widely reported, and credulously considered – was not in fact reliable. The claim being made was limited in one crucial respect. It was simply not true.

It did not take long for critics to highlight the many restrictive definitions the authors had used in their analyses (Smith, 2016). They had focused on hurricanes, yes, but this meant they had opted to ignore tropical storms, which are similar and just as deadly. They considered Atlantic hurricanes but not Pacific ones, without an obvious reason for geo-restriction. They confined themselves to hurricanes that made landfall, discounting those that remained offshore, even though offshore hurricanes are regularly just as lethal. (In 2009 the offshore Hurricane Bill became so famous that ten thousand sightseers gathered along the coast of Maine in order to look at it, despite coastguard warnings telling them to stay away. A single wave washed twenty of them into the ocean. Clearly, people didn't give Hurricane Bill much respect, even though it had a male name.)

Most strikingly of all, the authors only considered fatalities within the United States, even though many of the hurricanes they studied affected neighbouring nations too. In 1980 Hurricane Allen killed 269 people in several Caribbean countries, but only two people in the USA. The authors duly recorded Hurricane Allen as one of the *safest in* their dataset. After all, it killed only two Americans.

The authors had made a series of peculiar choices. They could have included Hurricane Bill, but they opted not to. They could have considered all the fatalities resulting from Hurricane Allen, but they didn't. They could have looked at tropical storms, and events in the Pacific, but they decided to exclude them. Instead they shaved and slimmed their dataset in a way that produced a profound effect on their finding. Their choices made female hurricanes seem more lethal and male ones appear safer.

Statisticians refer to this type of thing as the 'torturing' of a dataset. Such mistreatment of data is as nasty as it sounds. Moreover, as is generally well known nowadays, there are two significant problems with torture. Firstly, it is unethical: it is not morally correct to do it. And secondly, it is ineffective: what gets blurted out under duress might sound good, but, more often than not, it is just not true.

Many researchers have attempted to reproduce the finding that female hurricanes are deadlier than male ones. None have been able to do so. In fact, every single subsequent study has shown that male hurricanes are just as deadly as their female equivalents (Smith, 2016). It turns out that people don't attribute gender stereotypes to the weather after all. They treat girl hurricanes with just as much respect as any other.

Despite making global headlines and remaining on the record as a result of peer-reviewed published research, the finding that female hurricanes are deadlier than male ones is demonstrably specious. Despite many attempts, the result cannot be replicated.

'Fundamentally squishy psychobabble'

It never looks good for psychology when its research findings turn out to be false. It stains the field's image. This is unfortunate, because psychologists are normally quite image-conscious. Professional bodies devote enormous energy to outreach and media work. They develop detailed strategies to optimize psychology's 'footprint' and 'impact'. They produce packed portfolios of press releases, breathlessly pitching even very vague research summaries as potential news stories. University psychology departments often operate like press offices, feeding stories directly to news outlets, sometimes on a weekly basis. The total collective effort expended in enhancing the field's media profile is little short of monumental.

Presumably, these endeavours are intended to avoid headlines like the following:

Study Reveals That a Lot of Psychology Research Really Is Just 'Psycho-babble'.

Psychobabble is certainly not a term that psychologists would choose to describe their work. But it was the word chosen for them by the UK's *Independent* newspaper in August 2015, when covering a just-published critique of the quality of psychology research (Connor, 2015). The *Independent* succinctly summarized the critique's main finding:

Psychology has long been the butt of jokes about its deep insight into the human mind...and now a study has revealed that much of its published research really is psycho-babble.

For 'psychobabble' to be mentioned once in a national newspaper is bad enough. For the term to be used repeatedly is certainly not in any psychologist's media strategy.

This psychobabble conclusion was widely reported. The *Guardian* described the finding as 'a bleak verdict on the validity of psychology experiment results' (Sample, 2015). The *New York Times* felt it 'confirmed the worst fears of scientists' regarding the state of psychology (Carey, 2015). According to the *Washington Post*, it confirmed that much of what is published in psychology journals is 'fundamentally squishy' (Achenbach, 2015). Only a modicum of French-language proficiency is required to interpret the verdict of *Le Monde*: 'La psychologie est-elle en crise?' (Larousserie, 2016). By not mentioning psychology in its headline the *Huffington Post* (Reid, 2016) managed to hide the field's blushes (but only just):

Scientific Study Proves Scientific Studies Can't Prove Anything.

Psychologists had been whispering about such a crisis for decades. But now the idea that psychology research is not replicable was about to go viral.

The excitement was stimulated by work produced by the Open Science Collaboration, an international group of over two hundred researchers led by University of Virginia psychology professor Brian Nosek. The group published an extensive and damning report in the elite academic journal *Science* (Open Science Collaboration, 2015). In their programme, they attempted to re-conduct one hundred published psychology experiments. Over 60 per cent of the attempted replications 'failed'.

In other words, the results of the re-conducted studies bore little or no similarity to those of the originals they were based on. In psychology, as in any other science, this is a real problem. When the same study done twice produces two different outcomes, then one thing is for sure: both cannot be right. When 60 per cent of studies are like this, the accuracy of the entire field is called into question. As a column in *Nature* (Baker, 2015), the world's leading scientific journal, bluntly put the point:

> Don't trust everything you read in the psychology literature. In fact, two thirds of it should probably be distrusted.

The problem of course is this: how do you know that the study you are interested in – or indeed any particular psychology research study that you happen to encounter – is one of the trustworthy few and not one of the untrustable majority?

It is little wonder that journalists were interested. Psychology research affects our lives in so many ways. It is consulted in the design of everything from road safety campaigns, to anti-litter initiatives, to health promotion programmes, to educational curricula, to psychometric tests, to household products, to advertising. Politicians cite psychology research when debating policies about childcare, or poverty, or social exclusion, or environmental protection. Even disputes about major constitutional issues, such as whether to extend access to abortion or to lower the voting age, will be influenced by what psychology research has revealed about these issues.

Psychology relates to so much of daily life that its research has universal resonance. If it is actually true that 60 per cent of psychological science is not replicable – if psychology research is indeed 'fundamentally squishy psychobabble' – then of course the world's media will report this. They have a moral obligation to do so.

According to a famous quotation often attributed to Albert Einstein, insanity can be defined as doing the same thing over and over again and expecting different results. In psychology, such a scenario is the standard way of things. When psychologists conduct the same study over and over again, it would be reasonable to expect *different* results each time, given the historical pattern to date. In Einsteinian terms, it is the confident expectation that a replication might *succeed* that should rightly be considered madness.

The recurrence of non-replication

It would be wrong to think that psychology's replication crisis is new news. In fact, it is very much old news. The Open Science Collaboration itself was the culmination of decades of concern, the outpouring of extreme frustration among an increasingly disaffected subgroup of the psychology research community.

Psychology is often called the 'science of behaviour'. These days, we take behaviour to include people's actions, thoughts, and feelings. Unlike other sciences, psychology is not just a method of inquiry, it is also a field of human activity that itself can be inquired into. Not only can psychologists examine how people in general act, think, and feel, they can also examine how they themselves act, think, and feel. Psychology is itself a behaviour. Therefore, psychology can – and possibly should – study itself.

In historical terms, the use of logic and evidence to uncover knowledge is a relatively modern thing. Before science, and for many centuries, people had little choice but to draw knowledge from authority figures, superstition, or popular consensus. Any claim, no matter how weak, stood a good chance of being believed. People thought that diseases were transmitted through foul-smelling vapours, that all living beings were animated by a spirit-like force, and that California was an island. Science changed all this. It elevated the status of logic and evidence in public discourse. It attached value to validated accuracy. Intellectually, more than technologically, science altered the course of humanity.

Throughout the world, a scientific culture developed during the sixteenth, seventeenth, and eighteenth centuries. Central to science was the idea of objective verifiability. If evidence was to be valued, it had to be verified. If independent observers could not agree that the evidence stacked up, then even once-promising empirical claims would have to be discarded. Verification helped bolster the knowledge base. Falsification – where researchers deliberately sought out evidence that might *disprove* a given claim – bolstered the knowledge base even more. The core tenet was that science not only established the facts, it sought to check them and then to re-check them, and then to re-check them again.

By most accounts, psychology became a mainstream science during the middle of the nineteenth century. This means that psychological research has been conducted for over one hundred and fifty years. Literally millions of papers have been published. Given science's emphasis on objective verifiability, we might reasonably assume that a huge volume of this work has involved the checking and re-checking of previous findings. However, this is simply not the case. According to one analysis, only 1 per cent of papers published in the top one hundred psychology journals *during the last one hundred years* have been replications (Makel, Plucker, and Hegarty, 2012). Replications in psychology are not just uncommon. They are almost unheard of.

Nonetheless, given the overall volume of research during this period, 1 per cent still amounts to thousands of studies. Therefore, it should be possible to evaluate the effectiveness of these replications. Computations of psychology's replication rate have thrown up two main conclusions.

Firstly, quite often published replications confirm the results of the original studies on which they are based. But secondly, a worrying pattern emerges when it comes to study authorship: replications are likely to corroborate previous findings when they are performed by *the same researchers who conducted the original studies*. They are much less likely to do so when conducted by independent investigators (Makel, Plucker, and Hegarty, 2012).

In other words, when psychologists check their own homework, they tend to do very well. But when someone else checks it for them, they tend to flunk.

All this raises two important conclusions regarding replication in psychology. The first is extremely disappointing, the second extremely worrying. The disappointing conclusion is as follows. Despite the importance of objective verification and replicability in science, psychology clearly lacks a replication culture. Psychology can hardly convince audiences that its findings are replicable when custom and practice is to avoid even *attempting* to replicate research.

The worrying conclusion is that this reluctance to replicate may be hiding a much bigger problem. The link between study authorship and successful replicability – where replications 'succeed' when conducted by friends but 'fail' when conducted by strangers – raises questions about scientific objectivity. It closely resembles what would likely happen if researchers deliberately biased their studies to produce outcomes they particularly wanted.

For many years, this was psychology's dirty little secret. This is no longer the case. That is, it is no longer a secret.

The prehistory of psychology's replication crisis

The idea that psychology might have a problem with replication is as old as psychology itself. As soon as psychology emerged during the late nineteenth century, there was concern that what was taught might not be reliable.

The first psychology departments in Europe and North America were effectively repurposed religious philosophy departments: contemplation of the soul had gradually morphed into research on the mind (Fuchs, 2000). The Victorian fascination with psychics and séances had a strong influence, even on the most rarefied academic institutions. In the late 1800s Harvard University's psychology department listed a Ouija board in its inventory of laboratory equipment (Coon, 1992). The tenuousness of such subject matter proved disconcerting to many.

Leading psychologists feared ridicule were psychology to become sucked into the world of esoteric ephemera. Many mediums whose powers were praised with earnest sincerity by eminent scholars turned out to be charlatans (Coon, 1992). So extensive were the problems, psychologists quickly recognized that unguarded observations were too naïve to be relied upon for research purposes. Formal scepticism was required. As a result, several now-standard techniques for strengthening research – including blinding and counterbalancing – were developed during this period (O'Keeffe and Wiseman, 2005).

The late nineteenth century also saw the emergence of Freudianism, and with it the psychoanalytic school of psychology. While psychoanalysis proved extremely popular, it also attracted suspicion from the scientifically minded. A core tenet of psychoanalysis was the claim that the unconscious – an entity that by its very definition was unobservable – truly existed. According to critics, such a theory was the opposite of science. It completely disregarded the principle of objective verifiability. It generated a quagmire of open-ended prognostications. Concerns about whether psychoanalysis could ever produce replicable findings quickly became widespread.

By the twentieth century psychologists had become very exercised about rigour. There emerged a prevailing view that they should only concern themselves with observable – and thus replicable – phenomena. Figures such as Fechner and Wundt developed the field of psychophysics, which used standardized methods to examine the limits of human senses under laboratory conditions. By studying the dimmest lights that people could see or the smallest differences in volume that they could hear, the psychophysicists established that sense discrimination was mathematically predictable. Psychophysics was premised on the notion of replicability: data were informative only insofar as they were consistent (more or less) across all humans who were studied.

Knowledge arising from psychophysics could be helpful in certain contexts, but its applicability was limited. Its focus on exactitude served to restrict the range of subjects that could be studied. Psychology was much more than psychophysics, and it required much more powerful approaches.

Many psychologists formed the opinion that a better way to study human experience was to focus on externally observable behaviours. This led to the emergence of behaviourism, a theoretical view that dominated psychology for much of the twentieth century. Its rapid spread reflected pervasive unease about the replicability of psychological observations. Like the psychophysicists, the behaviourists strove to focus on quantifiable data and on whether what they saw in one human (or laboratory rat) could be generalized to others.

Psychophysics and behaviourism loom large in almost every telling of the history of psychology. Both emerged in response to anxieties about non-replicability. These fields spearheaded an emphasis on scientific rigour that now exists across psychological research, but they also highlight the fact that non-replicability is by no means an exclusively modern concern.

The problem of Rampant Methodological Flexibility

This exponential growth in scientific psychology brings its own challenges. At any given moment, thousands of investigators conduct tens of thousands of studies, the results of only some of which will eventually see the light of day in print. There is no one standard way to conduct psychology research. Much of what is planned and executed is devised by the researcher concerned and so is subject to whatever choices or preferences the researcher has at the time. There has emerged in psychology a culture of Rampant Methodological Flexibility.

Rampant Methodological Flexibility makes it extremely difficult to compare different studies. While two researchers in separate laboratories might conduct research on the same topic, their methods will inevitably deviate. Whether one study truly replicates the other is unclear. If both studies produce similar results, the investigators can claim to be corroborating each other. But if the studies produce different results, the investigators can comfort themselves by attributing the divergence in findings to differences in research methods.

Researchers allow self-serving bias to addle their scientific brains. When faced with ambiguous results, it is part of human nature to think of your own study as being well designed, coherent, and worthwhile. Even the most conscientious researcher will be inclined to interpret data in a way that bolsters this self-flattering premise. Whether intentionally or unconsciously, researchers end up engaging in a practice now often called HARKing – 'hypothesizing-after-the-results-are-known' (Kerr, 1998). They employ logical fudges to make arbitrary sense of ambiguous findings.

Rampant Methodological Flexibility also allows researchers to make on-the-hoof decisions about how studies are conducted. They are free to nudge their results in desired directions. For example, if they check the data from their first twenty participants and see what they like, they can continue the study and produce the results they want. However, if they don't see what they like, they have options. They can discontinue the study and design a new one. Or they can modify the existing study and continue the research, having tweaked the procedure in a desired manner. Or they can decide, retrospectively, to rewrite their study 'prediction' so that their data no longer contradict their hypothesis. In effect, if they don't see what they like, they can choose to like what they see.

Ultimately, by the time the research is published in a scientific journal, all such discretionary methodological choices will be part of dim and distant history. Most will unlikely ever be reported. Readers of scientific journals are faced with thousands of studies that turned out the way they did because of self-serving researcher influence but that could have turned out differently if *circumstances* had been different. Readers only see studies that make it into press. They don't see the studies that were abandoned or those that would have taken place had the researchers made different design choices.

Astute psychologists noticed this problem many years ago. In 1967 Paul Meehl, a professor at the University of Minnesota, warned of 'eager-beaver researcher[s], undismayed by logic-of-science considerations', who wend their way through strings of ad hoc methodological adjustments, publishing study after study but producing findings that simply cannot be relied upon (Meehl, 1967). Meehl saw this problem as extending far beyond self-serving bias among investigators. He described a structural problem that affected psychology as a whole. Psychology did not just *tolerate* methodological flexibility; it actively rewarded and encouraged it. His observations were published in the niche academic journal *Philosophy of Science* half a century ago. Nearly fifty years later the Open Science Collaboration – and the world's media – were reporting that not that much had changed.

The modern replication crisis

At the turn of the twenty-first century murmurs about psychology's structural replication problem became a burgeoning hubbub of intrigue and ignominy. In some ways psychology had changed. The field had grown to overlap with sister disciplines such as sociology and biology. Its subject-area remit widened to incorporate a range of social and cultural issues, as well as some hard-core neuroscience. But in other respects psychology had stayed the same. The new topics and approaches did little to alter the problems caused by Rampant Methodological Flexibility.

Indeed, in many ways they had made them worse. For example, brain imaging researchers discovered they were getting their sums wrong. Their conventional statistical approaches grossly inflated results, leading to entirely spurious findings in many cases (Vul et al., 2009). The problems seen in imaging studies also hampered research on event-related potentials, another technology used to measure brain activity (Luck and Gaspelin, 2017). Around half of all papers in top-tier neuroscience journals were found to report statistical interactions ineptly, generating waves of false conclusions about their meaning (Nieuwenhuis, Forstmann, and Wagenmakers, 2011).

Contributing to these problems was the way in which neuroscientists apply high-powered statistical procedures to low-powered tiny studies. It is very much a case of using sledgehammers to crack nuts. Simply put, most brain imaging studies use samples too small to support confident statistical conclusions. According to one assessment, brain imaging studies are so minuscule they can be expected to find no more than eight out of every one hundred real effects that exist (Button et al., 2013). This means that, for every hundred effects described in the literature, as many as ninety-two are liable to be flaky. They are likely to belong to the category of 'false positives': outcomes that *look* like findings but that aren't really findings at all, just freakish statistical patterns produced by chance.

In the years leading up to the Open Science Collaboration, several other investigations were casting different kinds of doubt on psychology. Questionable research practices were shown to be common. In one survey the authors of over 140 studies appearing in major psychology journals were asked to share their datasets (Wicherts, Bakker, and Molenaar, 2011). Even though all had previously signed agreements to do so (a requirement for research to be published), the majority now refused. Researchers who had reported the most tenuous findings were most likely to decline, suggesting an awareness of weaknesses in their work.

Statisticians used modern analytic techniques to gather direct evidence of the problem of Rampant Methodological Flexibility. Their approach was to scrutinize large swathes of published psychology research and to compare its findings to statistical models of what would have been produced had the science been truly unbiased. This work revealed an overall statistical pattern of results that was so improbable, it implied a large proportion of published findings in psychology are either cherry-picked, manipulated, erroneous – or simply made up (Leggett et al., 2013).

The prospect of fraud should not be dismissed lightly. Survey results suggest that around one in seven psychologists are directly aware of colleagues who have falsified data (Fanelli, 2009), and a number of particularly audacious fakery cases, some in very high-profile research groups, made international news headlines in the early 2000s (Estes, 2012).

But not all factors leading to non-replicability relate to researcher mis-behaviour. Some reflect adverse attitudes among the intended gatekeep-ers of scientific standards. The fact that replications appear so rarely in psychology journals is only partly due to the lack of studies. To a greater extent, it reflects the reluctance of editors to consider such papers for publication.

In late 2015 the Open Science Collaboration published the long-awaited findings of its 'Reproducibility Project'. Over a three-year period lead author Brian Nosek had persuaded 269 colleagues from around the world to join him in attempting to replicate one hundred studies from major psychology journals. To focus on a fair selection of research, Nosek randomly drew studies from a single year (2008) and from three top-tier journals, namely: *Psychological Science*, the *Journal of Experimental Psychology: Learning, Memory, and Cognition*, and the *Journal of Personality and Social Psychology*. He then categorized the selected studies as either social psychology (i.e., those from the *Journal of Personality and Social Psychology* and half of those from *Psychological Science*) or cognitive psy-chology (i.e., those from the *Journal of Experimental Psychology* and the remaining ones from *Psychological Science*).

Detailed protocols were established for constructing and conducting the replications. Investigators were required to contact the original authors for study materials and to invite them to review the replication methods. To avoid the risk of HARKing, investigators also had to publicize their methods before commencing their replication. After all the replications were completed, their results were compared with those of the originals. Nosek and his colleagues considered two types of comparison. First, they looked at statistical significance – in other words, whether the replication yielded evidence in support of the original research prediction. Then they looked at the size of the statistical effects – in other words, whether the replicated result was similar to the original not just in trend but also in intensity or impact.

As we know, the picture was wholly underwhelming. While 97 per cent of the original studies had produced statistically significant find-ings, only 36 per cent of the replications did so. Replicability in social psychology seemed particularly poor. Only 23 per cent of studies from the *Journal of Personality and Social Psychology* were successfully replicated, along with just 29 per cent of the social psychology studies from *Psychological Science*. While cognitive psychology proved more replicable, its record was still far from stellar. Only around half of these replications were successful (48 per cent of those from the *Journal of Experimental Psychology* and 53 per cent of cognitive studies from *Psychological Science*). Nosek and his colleagues then performed computations to calculate a score, on a scale of 0 to 1, that represented the effect size of the various studies. The median effect size of the original studies was 0.403; that of the replications was just 0.197. This showed that even where studies seemed replicable, replications produced much weaker findings than the original research.

The picture is clear. Psychology has a problem with replication. This problem is not restricted to social psychology, which is often said to be difficult to study scientifically. It also affects cognitive psychology and neuroscience, both said to be very scientific. The problem is not confined to research in obscure journals with dubious editorial standards; it emerges throughout top-tier journals also. And it is not just the result of self-serving bias among researchers; it is exacerbated by editorial practices that encourage Rampant Methodological Flexibility and discourage the conducting of replications.

The problem takes two main forms. Firstly, psychology flagrantly lacks one of the key hallmarks of science – a *culture* of replication. And secondly, large numbers of psychology studies fail the basic test of science – under the challenge of replication, they fail.

Non-replicable findings are everywhere

Psychologists like the term confirmation bias. It describes how people always try to prove themselves right. When judging an unproven proposition that they like, most people are selective when it comes to the evidence they look for. They are also selective in interpreting the evidence they find.

People are rarely truly objective. They over-focus on information that supports their prior beliefs, and they generously interpret ambiguous information as if it was clearly in their favour. In short, they engage in logical fraud. They pretend the evidence is more conclusive than it actually is, and they congratulate themselves on their perceptiveness in having been proved 'right'. To be fair, most people do this somewhat unconsciously, and this is what is properly referred to as *confirmation bias.* When it is done consciously, it is good old-fashioned cheating.

Psychologists have conducted thousands of research studies on confirmation bias, so you would think they would know lots about it by now. They should certainly be familiar with the issue. This is because as well as studying confirmation bias, they also succumb to it. When poor quality psychology research is published or cited widely, it is nearly always because of confirmation bias.

Research is considered much more plausible when researchers, reviewers, editors, and ultimately readers believe that the findings 'make sense'. As such, results that support a commonly held view, or a fashionable theory, or an obscure principle held dear by technical specialists are much less likely to be intensively scrutinized. Less attention is paid to the methodologies used to produce them.

It is very common for weak research papers to receive generous evaluations because editors and reviewers become swayed by the expectation that studies should produce 'logical' results. Female hurricanes are more deadly? Well it makes sense; after all, implicit sexism is an important social problem. Psychology journals accumulate many papers because of such confirmation biases, perhaps more than are accepted because of compellingly sound science.

Take, for example, a 2011 study in the journal *Applied Animal Behaviour Science*, which claimed to show that dogs are four times more likely to bite other dogs when they are walked by a man than when they are walked by a woman (Řezáč et al., 2011). The implication is that male dog-walkers, by virtue of being men, somehow transmit their masculine aggression to the dogs they walk. It was widely reported that the study involved the observation of nearly two thousand dogs and their owners: the Discovery Channel referred to it as 'one of the world's largest studies of dogs on walks' (Viegas, 2011). However, the research was weak in several respects.

Firstly, it was an observational study in which, across several weeks, student research assistants lurked in public parks with clipboards and watched people walking their dogs. Whether the same dogs and walkers were included on multiple occasions was not accounted for. Secondly, the study was cross-sectional, so it is impossible to infer cause-and-effect relationships. Maybe more aggressive dogs are walked by men precisely *because* they are more aggressive. In other words, maybe there is a tendency among dog-owning families to ask their menfolk to walk the more troublesome canines. Thirdly, the data were analysed in a very simplistic way. The authors used bivariate instead of multivariate techniques. This means that they did not seek to statistically control for multiple other potential causes of dog aggression, such as dog size, when analysing gender differences. And finally, the study was far from one of the world's largest studies of dogs on walks. In fact, it might have been one of the smallest. For sure, around two thousand dogs were observed in the research as a whole. However, only twenty-eight were seen biting (this was the authors' working definition of dog aggression). As such, only this subgroup of twenty-eight – effectively comprising a mini-study within the overall research project – could be used to investigate the impact of gender-of-walker on dog aggression. By any standard, this is a fatally insufficient sample. It seems that confirmation bias drove the publication of this study and its coverage in the news media. The study's many weaknesses, including the tiny sample size underlying its headline finding, were overlooked. To readers, the idea that male dog-walkers make their dogs more aggressive seemed to 'make sense'. It conformed to a cultural stereotype that depicts men as inherently more violent than women.

In recent years an increasing number of initially attention-grabbing psychology studies have ended up foundering due to this confirmation bias problem. Many relate to claims about personal fate and well-being. In one very prominent example, researchers asserted that adopting a 'power pose' – standing with your hands on your hips and your elbows pointing outwards, like a comic-book superhero – produces several immediate benefits, including enhanced feelings of competence, increased willingness to take risks, lower levels of the stress-hormone cortisol, and higher levels of testosterone (Carney, Cuddy, and Yap, 2010). One of the authors delivered a TED talk based on the research, which has now been viewed over 40 million times. In it she claims that power poses are so impactful, people who use them 'can significantly change the outcomes of their life' in just two minutes (Cuddy, 2012).

In some ways this seems like quite an implausible notion, one that few people would have anticipated in advance, and so immune from confirmation bias per se. However, the idea that people can exert agency over their own destiny is a very strong cultural belief (Yarritu, Matute, and Vadillo, 2014), as is the notion that physical posture is associated with inner psychological character (Gilman, 2014). This, coupled with the Ivy League affiliations of the study authors and the esteem of the publishing journal, created perfect conditions for confirmation bias. In short, readers of the research quickly concluded that, all things considered, the findings 'made sense'. The problem, of course, is that they really didn't. The study was based on a sample of just forty-two participants. When independent investigators conducted replications based on samples of two and three hundred, they found no power pose effects whatsoever on any of the alleged outcome variables (Garrison, Tang, and Schmeichel, 2016; Ranehill et al., 2015). The original power pose findings have been thoroughly debunked, so much so that the lead author published a statement on her personal website repudiating them in forthright terms (Carney, 2016). It turns out it takes more than two minutes of standing like a superhero to improve your life outcomes.

Similar doubts have surfaced concerning other research linking simple behaviours to complex consequences. These include a famous study claiming to show that moving your facial muscles into the shape of a smile – such as when you hold a pencil sideways in your mouth – causes neurological feedback that engenders feelings of happiness (Strack, Martin, and Stepper, 1988). Notwithstanding the extent to which this study is cited in introductory psychology textbooks, it turns out the reported effect cannot be replicated (Wagenmakers et al., 2016).

A comparable controversy surrounds research on the concept of 'ego depletion', an effect in which people's willpower is shown to be a finite resource that declines the more it is used (Baumeister et al., 1998). The original study has been extensively cited, and several researchers have claimed to have demonstrated ego depletion in different experiments. However, critics have long questioned whether it really exists. One analysis of over a hundred ego depletion experiments showed their combined statistical result to be effectively zero (Carter et al., 2015). A subsequent literature review found that ego depletion studies show a bias towards false-positive (i.e., random) effects (Schimmack, 2016). More damningly, a direct replication of the original study, conducted by independent investigators using a much larger sample, found no evidence for ego depletion at all (Hagger et al., 2016).

In another famous study, researchers claimed that participants walk more slowly after being presented with words like 'bingo' and 'Florida', words that the study's authors argue plant stereotypes of the elderly in participants' minds (Bargh, Chen, and Burrows, 1996). The finding stumbled and fell upon replication (Doyen et al., 2012).

Other researchers have claimed to show that getting people to hold hot cups of coffee makes them more likely to rate others as generous and caring (Williams and Bargh, 2008). The effect implied that making people physically warm also makes them interpersonally warm. A replication based on a much larger sample was sufficient to pour cold water on that conclusion (Lynott et al., 2014).

A related line of research has attempted to link cleanliness with morality. While attracting positive attention for many years, the field is in disarray now that replications have shown the effects to be untrustworthy. One experiment claimed that handwashing makes people more forgiving (in other words, that hygienic purity engenders a sense of moral purity; Schnall, Benton, and Harvey, 2008). Another claimed that threatening people's sense of morality makes them seek out ways to cleanse themselves (in other words, that moral purity engenders a craving for hygienic purity; Zhong and Liljenquist, 2006). Both widely cited experiments have been subjected to robust replication attempts (Earp et al., 2014; Johnson, Cheung, and Donnellan, 2014). In both cases the original findings have been washed away.

Research that cannot be replicated comes in all shapes and sizes. A study suggesting that people pursue professions that remind them of their own name (e.g., women named Laura are more likely to become lawyers, while men named Dennis are more likely to be dentists; Pelham, Mirenberg, and Jones, 2002) is cited in several introductory psychology textbooks. However, according to a replication, the effect is actually spurious (Simonsohn, 2011). A related finding that people are more inclined to marry someone whose name is similar to their own (making Josephine more likely to plight her troth to Joseph; Jones et al., 2004) also disappeared when the relevant work was replicated (Simonsohn, 2011).

An influential study of day-old babies, which was supposed to show that baby boys liked staring at mechanical mobiles whereas baby girls preferred to gaze at human faces (Connellan et al., 2000), saw its finding vanish in a replication attempt (Leeb and Rejskind, 2004). A replication of another classic baby study (Meltzoff and Moore, 1977) showed that newborns do not, after all, possess the ability to imitate facial expressions of adults (Oostenbroek et al., 2016).

Seeing a picture of a pair of human eyes (e.g., as part of a sign) was originally reported to make people engage in more prosocial behaviour (Bateson, Nettle, and Roberts, 2006), and at least one police force adopted the idea when designing its anti-crime posters. But an independent attempt to replicate the finding in a larger sample found that, actually, no such effect occurs (Carbon and Hesslinger, 2011).

This is just a small selection of non-replicable research in psychology. In the main, these studies are well known: most are regularly cited in textbooks, and many are considered classics. When replications cause the core of the undergraduate curriculum to collapse like a house of cards, it becomes clear that psychology has a serious problem.

It should be recalled that replications make up only 1 per cent of published psychology research. For every high-profile study that attracts a replication, there are hundreds of others, most of them banal, whose replicability remains unknown. The true extent of non-replicability in psychology is a matter of speculation. But we can be sure that it is likely to be huge.

Why is psychology at risk?

So why is psychology's replicability problem so chronic? One way to answer this is to consider the factors that make spurious findings more likely. For example, we know that small studies are more likely to produce false-positive results than large studies. Therefore, if a field is dominated by small studies, its rate of false-positive results will be high. Similarly, we know that fields that study small effects (e.g., subtle group differences that are hard to notice without the aid of fine-grained statistical information) are more likely to produce false positives than fields that study large effects. Therefore, if a field is dominated by the study of small effects, then *its* rate of false positives will be high.

We know that taking a scattergun approach to statistical analysis – testing everything that can be tested in the hope of finding something statistically interesting – also heightens the likelihood of false-positive results. Therefore, if a field is characterized by such scattergun approaches, then it too can be expected to produce very high rates of false-positive findings.

And fourthly, if research in a given field is conducted with a high degree of methodological flexibility, wide-ranging analytic approaches, and inconsistent definitions of key terms, then *its* rate of false-positive findings can also be expected to be very high. Therefore, a field whose norm is Rampant Methodological Flexibility can be expected to produce a torrent of spurious, unreliable, and non-replicable findings.

The bad news is that psychology possesses not just some but *all* of these characteristics. Indeed, in addition, psychology possesses several other features that undermine research accuracy (such as: a proneness to study fashionable, but not well-understood, subject matter; a proneness to study topics about which researchers may hold personal biases or prejudices; and a culture of hectic competition among researchers to be the first to publish findings on a given topic).

One medical statistician has attempted to quantify the impact of all such factors in order to compute the general likelihood of spurious findings in the biological and health sciences (including psychology). According to his analysis, the odds of false positives are greater than fifty-fifty. Therefore, as he puts it, 'most claimed research findings are false' (Ioannidis, 2005).

Other statisticians have noted that studies published in scientific journals represent only a subset of those that have actually been conducted in the world. Scientists often commence studies that they eventually choose to abandon or abort. This usually happens when they conclude that a study's finding is trivial, nondescript, or otherwise boring. Quite often it is because the finding is not so much a finding as it is a non-finding: the predicted effect didn't materialize, or the anticipated group difference just wasn't there. In other words, the researcher's original conjecture was wrong. Rather than publish such a non-finding, the researcher discards the study, moves on, and invests their effort in designing new research.

The problem is that this serves to filter banal findings from the published literature. Readers who study the contents of journals are left with a skewed impression of the evidence. They will see plenty of statistically significant findings (the majority of which, according to Ioannidis, will be spurious). But they will *not* see those studies on the same topics that were abandoned because the researcher found nothing of interest in the data. In other words, published research on a given topic is difficult – if not impossible – to interpret unless you also have full information about the *unpublished* research that has been conducted on the same topic. This quandary was noted very long ago by an American psychologist called Robert Rosenthal. In reference to the way in which researchers stored their old unwanted papers at the time, he described it as the 'file-drawer problem' (Rosenthal, 1979).

Nowadays, the extent to which published research represents a biased selection of all research can be inferred statistically. As we will see in Chapter 4, psychology research typically determines the importance of findings based on statistical computations of probability. If the data collected are found to be improbable (allowing for various assumptions), then the finding is considered to be of interest. More specifically, if the statistical test shows this improbability to be less than one in twenty (i.e., if its 'p-value' is less than 0.05), then the result is declared 'statistically significant'.

Strictly speaking, as psychology research looks at natural phenomena, the probabilities computed for the data in such studies should reflect patterns seen in nature. The numbers should be spread like raindrops on a window pane or leaves on a tree: any vividly recurring pattern or exact right-angles would suggest human interference. However, when statisticians scrutinize the shape of psychology findings, they find many curious contours. The patterns deviate from what should be seen in a fair selection of research.

For example, statisticians have observed an inexplicable spike in p-values lying just below the magic 0.05 level (Leggett et al., 2013). In reality, there is no reason for the probability of naturally occurring data to cluster around a point just below 0.05 (or indeed around any other number). The fact that psychology data do this suggests *either* that they have been analysed in idiosyncratic ways in order to contrive this very outcome *or* that the studies themselves are highly selected and do not represent a fair sample of what has been done. Actually, both scenarios could be true.

Perhaps the biggest factor producing so many spurious findings is Rampant Methodological Flexibility. Unlike many other sciences, psychology lacks a standard experimental procedure. There is no single approved approach to research. Even studies that use comparable methods to examine similar questions will end up distinct from one other. Individual researchers are free to customize their research designs, alter their procedures, tweak their analytic practices, redefine their terms, and even retrospectively rewrite their hypotheses. This level of flexibility helps make psychology an adaptable field that can be tailored to any topic or research circumstance.

However, Rampant Methodological Flexibility also opens the door, extremely widely, to confirmation bias. It allows dishonest researchers to distort studies to suit their own needs, corroborate their own predictions, or otherwise make themselves look good. Honest researchers can also be tempted to do these things, albeit unintentionally. Whether researchers are dishonest or honest makes little difference to replicability. By facilitating confirmation bias, Rampant Methodological Flexibility makes it almost certain that psychology will produce plenty of spurious findings that simply can't be replicated.

The problems are compounded by commonplace scholarly weaknesses. Psychology is full of weak statistical practices, weak research methods, and weak theory. Weak statistical practices include a near-obsessive focus on probability values produced by statistical tests. As we will see in Chapter 4, researchers in psychology regularly fail to apply statistical tests correctly, do not fully understand the nature or purpose of such tests, and habitually misinterpret their results. Further, many psychologists – if not most – make unsound distinctions between findings based on whether the associated p-values are below or above the 0.05 threshold. This near-obsession with categorizing p-values leads to several erratic approaches to analysis (such as those that produce a bulge in ps just below 0.05). Rampant Methodological Flexibility allows researchers to try any number of statistical approaches in an effort to test and retest the same dataset to hunt for a suitable p. Whenever p falls below 0.05, an expedient researcher can choose there and then to cease the study, desist from any further analysis, and punch the air in celebration. Indeed, the culture of celebrating statistical significance as if it were some kind of victory (leading many a disappointed student to ask their supervisors 'What went wrong?' when their p is too high) is itself a serious problem.

The belief that the only good finding is a statistically significant one pressurizes researchers to exploit methodological flexibility in ways that undermine objectivity. Needing to resort to idiosyncratic statistical analyses in order to shake a significant *p*-value from an otherwise moribund dataset virtually guarantees that the finding so produced will be non-replicable.

Psychology's weak research methods include a decades-long and incorrigibly persistent reliance on small study samples. As mentioned above, sample sizes in brain imaging studies are, on average, so small that only eight out of every hundred real effects can reliably be observed (Button et al., 2013). In psychology as a whole, average sample sizes are large enough to detect only twenty-four effects out of every hundred (Smaldino and McElreath, 2016). Given that the vast majority of research papers report statistically significant findings, this could suggest that between 80 and 90 per cent of them are likely to be false positives (we will consider this matter further in Chapter 5). The fact that psychologists are often sanguine about the sample-size problem may reflect a generally casual attitude to the broader issue of research methods.

While research methods training is a core requirement of psychology degree education, it is generally considered socially acceptable among psychologists (especially among applied practitioners) to be somewhat disregarding of their importance. Some movements within psychology have adopted scepticism toward quantitative methods as their badge of honour. Subfields such as critical psychology and qualitative psychology not only argue that statistical methods are inessential, some voices can be heard claiming they are so ill-suited to the study of human behaviour as to be a blight on the social sciences. Given this culture, perhaps it is no surprise that some psychologists fail to prioritize statistical competence when attempting to conduct research.

Weak theory is also problematic. In its pure form, science seeks to build future knowledge by methodically refining present knowledge. It inches ahead in incremental steps, plodding along, objectively oblivious to the limelight. In contrast, modern psychology is often keen to take large bounding strides forward, in bold and newsworthy ways. It seems that many top-tier psychology journals prioritize the publication of studies that can be deemed 'exciting' or 'quirky', giving high-profile platforms to such notions as power poses, priming, and precognition.

However, the more novel the conceptual content of research, the greater its risk of relying on unproven prior assertions. Rather than standing on the shoulders of giants, such studies attempt to levitate high in the sky in order to see more and further than other studies can. But the problem with wanting to levitate is that it is untenable, and probably impossible. Some distortion of reality will be required.

But is psychology's replication crisis overstated?

In any crisis, there are the people who ask 'Crisis? What crisis?' Psychology's replication crisis is no exception. Almost as soon as the Open Science Collaboration published the results of its reproducibility project, there were critics who complained that the work itself was an example of spurious research in psychology. In the *Times Higher Education*, social psychologists Wolfgang Stroebe and Miles Hewstone (2015) argued that the Collaboration's results were as untroubling as they were unsurprising. They noted that single studies are always weaker than the combined results of multiple studies, so therefore conducting single replications was not a good way of testing original experiments. Instead, they said, combined analyses of many studies (so-called meta-analyses) should be considered the gold standard. According to Stroebe and Hewstone, many such meta-analyses had shown social psychology findings to be reliable. They dismissed the Open Science Collaboration findings as 'meagre' and 'not very informative'.

Perhaps the highest-profile scepticism directed at the Open Science Collaboration came from Daniel Gilbert, a Harvard University social psychologist. His critique appeared in *Science*, the same journal in which the Open Science Collaboration results had been published (Gilbert et al., 2016). His main complaint was that the replications were not sufficiently similar to the original studies on which they were based. For example, a study on racism originally conducted in the United States was replicated in the Netherlands. Gilbert argued that sociopolitical differences between the two countries rendered the Dutch replication problematic, and so it was unsurprising its results failed to match those of the American original. According to Gilbert, when such methodological differences are factored into analyses, the replication rates observed by the Open Science Collaboration fall well within expectations.

Echoing Stroebe and Hewstone, Gilbert further argued that conducting single replications for each original study was an insufficient way to test for false positives. After all, if the two findings differ, it does not necessarily follow that the original study is the one that is spurious. Maybe the replication result is the outlier. By failing to allow for this possibility, Gilbert suggested, the Open Science Collaboration inadvertently tested the original studies against artificially high benchmarks. Gilbert also argued that the Collaboration failed to account for bias. He suggested that differences in methodologies between original studies and replications might reveal that the investigators who conducted the replications were biased (either explicitly or implicitly) and secretly willed their replications to fail.

The Open Science Collaboration published a response to Gilbert's criticisms (Anderson et al., 2016). They claimed that Gilbert's critique contained several inaccuracies and statistical misunderstandings. They reported that several of the replication methodologies Gilbert criticized were approved by the original studies' authors. They pointed out that Gilbert omitted to discuss a number of their findings that directly contradicted his critique. Finally, they turned Gilbert's point about study differences on its head. They agreed with Gilbert that the replication studies differed from the originals they were replicating. Given the fact that replications are conducted at different times and by different people, they noted that 'there is no such thing as exact replication'. As such, while Gilbert had argued that the Collaboration tested the original studies against artificially high expectations, the Collaboration responded by arguing that Gilbert had critiqued their replications against similarly inflated standards.

It is sometimes noted that problems with replication are not unique to psychology. This is certainly true. In medicine the development of rigorous protocols for randomized controlled trials (RCTs) was itself a response to widespread concerns about non-replicability in medical studies. Anxiety about the reproducibility of drug efficacy trials dates back to the 1950s (Bastian, 2016). As recently as 2012, an attempt to replicate the findings of fifty-four 'landmark' preclinical cancer research studies found that only six could in fact be replicated (Begley and Ellis, 2012). So psychology is not alone. But of course, multiple wrongs do not make a right. Just because other fields have replication crises of their own does not make psychology's any less grave.

Models or muddles?

In reality, concerns about psychology long predate the Open Science Collaboration. Whether or not their reproducibility project has short-comings, we know that the crisis exists. We *know* that psychology lacks a replication culture: replication accounts for just 1 per cent of published research. We *know* that replications are less likely to succeed when they are conducted independently. We *know* that Rampant Methodological Flexibility opens the door to confirmation bias. We *know* that psychology frequently touches on subject matter about which researchers will hold strong personal opinions, thereby straining their ability to remain objective. We *know* that the file-drawer problem exists and that, therefore, published studies overstate the commonness of effects. We *know* that sample sizes in psychology are statistically too tiny to inspire confidence in research. We *know* that effects in psychology are often very subtle and so risk flooding the literature with spurious false-positive findings.

We *know* that researchers wilfully engage in questionable research practices. We *know* that scattergun approaches to data analysis are the norm rather than the exception. We *know* that researchers can retrospectively rewrite their hypotheses once their results are known. We *know* that several landmark studies have been cited in textbooks for years, even though their findings cannot be reproduced by independent investigators. We *know* that editors and readers overlook methodological limitations when study results are deemed to 'make sense'. We *know*, in other words, that psychology is at high risk of suffering a perennial problem of non-reproducibility and that radical action is needed.

Psychology's problems with replication encompass multiple methodological crises. In the rest of this book, we will examine each in turn. Psychology has significant challenges with the task of measurement. If psychological concepts are not quantified properly, then findings based on such measures will be hard, if not impossible, to replicate. Psychology faces many problems with statistical analyses. Quite how best to apply statistical techniques to the types of data gathered in psychology research is a matter of constant debate. Psychology has a particular challenge with sampling. How exactly do psychologists ensure that the people they study, and the situations in which they study them, yield findings that can be generalized to humanity as a whole? When the findings are in, psychology then faces severe difficulties with interpretation. The temptation to fill explanatory gaps with subjective judgements creates more confusion than certainty.

Psychology also faces problems with its own coherence. In many ways the psychology community is more cacophony than chorus. It comprises several subgroups who view the world in very different and highly incompatible ways. Psychology's problems with replication don't require data. To produce conflicting conclusions, all you need is two psychologists. By adhering to different theoretical paradigms, psychologists can diametrically disagree on even the basic details of how human beings behave. Given this state of affairs, it is not difficult to see how diverging research findings might then also arise. In essence, psychology's replication crisis is bolstered by a fundamental paradigmatic crisis. It is to this problem of fragmentation that we now turn.

CHAPTER 2

'Black Is White': Psychology's Paradigmatic Crisis

What to believe?

The claim that we have recently entered a new era of 'fake news' is itself a piece of fake news. There is nothing new about news that is fake. The fear of false information appearing as news (such as female hurricanes being especially lethal) is a long-standing one. This is why we have aphorisms to warn us. *You can't believe everything you read in the newspapers. Don't let the truth get in the way of a good story.* Or as Oscar Wilde put it: *Journalism keeps us in touch with the ignorance of the community.*

But perhaps we can say that *fear* of fake news has intensified in recent years. Throughout the world, political dramas have drawn attention to the way in which deliberate misinformation can threaten democratic damage. The boundaries between fact and falsity become contested and occasionally breached. Public discourse is now sufficiently malleable to accommodate the idea that a 'provable falsehood' might instead be referred to as an 'alternative fact' (Blake, 2017), as though the distinction between true and false is little more than a partisan affectation. It has become reasonable for a frontline politician to claim that the voting public has 'had enough of experts', effectively to argue that expertise itself is an arbitrary quality of dubious value to real people (Mance, 2016).

Fake news, alternative facts, and expert-baiting are sometimes said to reflect the emergence of a 'post-truth' culture, where public debate is framed by appeals to emotion rather than appeals to fact. Truth is an appreciated virtue, but it is ill-defined. Everyone is invited to obtain their own version of truth; and as everyone believes themselves to be virtuous, collective gut instinct drives perceptions of reality.

In this world, claims (or research results) that 'feel true' or 'make sense' are more saleable and less likely to be scrutinized. But if another person's claim runs counter to your personal orthodoxy, you are free to invoke the burden of proof as a debating tactic. You can reject their claim indefinitely, until such time as they can produce definitive, irrefutable, compelling, and unquestionable evidence beyond the scope of dispute. Anything short of 100 per cent certainty can be taken as sufficient wiggle room to permit scepticism. But, of course, no assertion can conclusively be proven in this way; therefore, none can be disproven; therefore, all can be believed. This is the malaise of a post-truth culture.

The truth is out there. But we can't know where.

The ridiculing of facts and expertise makes it likelier that citizens will come to believe falsehoods. As such, it has been suggested that psychologists should look for ways to neutralize the effects of post-truth claims. In other words, some policymakers want psychologists to develop a 'vaccine' for fake news (van der Linden et al., 2017).

But before they do that, perhaps psychologists should first consider getting their own house in order. In a post-truth culture, opinion shapers allow emotions and aesthetics to colour their attitudes to evidence. Post-truth advocates ignore valid rebuttals of their claims, reject the merit of other people's expertise, and simply repeat their talking points even after they have been publicly debunked. Would it be fanciful to suggest that this sounds a lot like the behaviour of psychologists, at least from time to time?

Many psychologists appear to deny the replication crisis from positions of sentimental optimism, rather than reasoned analysis. They often remain loyal to favoured theories or practices long after their premises have been shown to be flawed. In some instances, psychologists reject the contributions of expert non-psychologists, even when their input is clearly relevant. And, judging by introductory psychology textbooks, it is common for psychological research findings to be cited again and again even after replications have shown the original result to be spurious.

For sure, there is something of a post-truth feel about much modern psychology.

Post-truth psychology

Sometimes psychology's fake news is of the old-fashioned type: a news story that, in the end, is shown to be fake. One of the most notorious examples was the 'news' that research had shown the mumps, measles, and rubella (MMR) vaccination to be implicated in the onset of autism. The core claim was made in a press conference that preceded the publication of an article in the medical journal, the *Lancet*. The article itself reported findings so surprising that several replications were eventually conducted by independent investigators. In the end, it was easy to demonstrate that the findings in the *Lancet* article were indeed highly questionable, but not before a great many parents around the world had become convinced that MMR was dangerous for their children.

However, the *Lancet* findings were not merely false-positive statistical flukes. They were not spurious results given artificial prominence by the file drawer problem. They were, in fact, the product of fraud. When journalists tracked down the original study participants, they found that much of the 'data' presented in the *Lancet* paper had been significantly altered by the authors (Deer, 2011). Some were concocted out of thin air. The authors were reprimanded by their regulatory body and struck off the medical register. The *Lancet* paper was formally retracted. In short, the claimed link between MMR vaccines and autism was shown to be fake news. Fake psychology news.

Examples of 'alternative facts' also abound in psychology. Some relate to egregious pseudoscience, such as the urban myth that human beings typically use just 10 per cent of their brains (Beyerstein, 1999). Others emerge from the replication crisis, or at least seem to. For example, as we saw in Chapter 1, assertions relating to facial feedback (such as the idea that holding a pen sideways in your mouth will make you happy), implicit egotism (such as the notion that people called Dennis are statistically more likely to become dentists), and embodied cognition (such as the claim that washing your hands will make you moralistic) would appear somewhat resistant to refutation. By their inclusion in textbooks long after they have been debunked, it could be said they meet the definition of alternative facts.

Many psychologists engage in expert-baiting, albeit often in subtle ways. Instead of denigrating experts of all kinds, they direct their fire at particular categories of expertise. One example is when they malign statistical expertise. Many social psychologists, while being adept at social psychology, could not be said to be experts in statistics or research methods. Yet some feel free to reject or attack the conclusions of expert statisticians and methodologists. This appears little different from when a career politician or business leader rejects the views of a climate scientist.

With regard to the replication crisis, such dismissals have been vituperative. One Ivy League psychologist complained that, far from being experts, these specialists were in fact 'shameless little bullies' (Bartlett, 2014). Another, a former president of the Association for Psychological Science, accused her critics of engaging in 'methodological terrorism' (Letzter, 2016). And beyond the replication crisis, as mentioned in Chapter 1, some subfields of psychology, such as critical psychology, are quite explicit in impugning the knowledge of statistically minded colleagues (Faux, 2014).

In a post-truth context claims to truth are never unassailable, and disputes about fact are never resolvable. The contention that psychology research can indeed be relied upon, and the ignoring of evidence that it cannot (or that much of it cannot), could be seen as a good example of post-truth psychology. Some psychologists say one thing, others say quite another thing, and eventually 'psychology is in crisis over whether it's in crisis' (Palmer, 2016).

But it is not unusual for some psychologists to say one thing while other psychologists contradict them. In fact, it is a standard feature of the field. Psychology is a field of many subfields. It comprises several schools of thought, or 'paradigms', each shaped around a particular view of why people think, feel, and act the way they do. The problem is that many of these views contradict each other. They are based on mutually incompatible assumptions about human nature. If you adhere to one school of psychology, then – whether you realize it or not – you inherently disagree with adherents of at least some other schools.

Psychologists do not need research or data to conclude that other psychologists are wrong. It might not be polite to say so, but psychology is theoretically sectarian.

On the one hand, on the other

Attempting to summarize developments in psychology over the preceding century, British psychologist Cyril Burt opined that

> psychology, having first bargained away its soul and then gone out of its mind, seems now, as it faces an untimely end, to have lost all consciousness. (Burt, 1962, p. 229)

Burt was commenting on the various ways in which psychologists, and others, had attempted to explain human nature. Psychology 'bargained away its soul' when it displaced the previously dominant religious view that humankind was best explained by invoking deities and spirituality. It went 'out of its mind' by centring on theories such as psychoanalysis, which linked all psychological experiences to the role of the unconscious mind, and attempted to build a complete theory of human psychology from the study of clinical patients with mental-health disorders. Finally, psychology 'lost all consciousness' by embracing behaviourism, a theoretical approach that focused on the role of environmental stimuli in determining human behaviour and dismissed the study of consciousness as unreliable.

Burt was writing in 1962, before the emergence of modern cognitive psychology, social psychology, biological psychology, humanistic psychology, and various other specialisms we see today. If he were writing now, his aphorism would need to meander for quite a bit longer. And it would need to take account of the fact that psychology has not just developed from one theoretical position to another across time. Nowadays psychology is capable of holding several such positions at the *same* time.

Take for example psychologists' explanations of the way in which people react to stressful situations. While there are several such accounts, two are useful to mention here. Some psychologists say that when faced with a challenge, we manifest a stress response based on an elaborate sequence of cognitions. We first ask ourselves how big the challenge is. If we conclude it is mild, then we are unlikely to get too bothered. However, if we perceive it to be significant, then further cognitions will flow: we will ask ourselves whether we can cope with such a significant challenge. This involves thinking about the resources we can draw upon (such as information, skills, money, or social support). If we conclude we have lots of resources we will remain physiologically calm. However, if we conclude there is a shortfall between our available resources and the demands posed by the challenge, then we will have a difficulty. It will make us distressed and so trigger a physiological response; our heart rate will rise, our breathing will quicken, and stress hormones will surge through our bloodstream.

The theory claims to explain why the same event can be experienced differently by different people. One person may perceive it to be challenging and to exceed resource capacity, and so find the event stressful.

A second person may perceive the same event as *not* challenging, and/or to fall *within* the capacity of available resources, and so find the event *not* stressful. In short, this theoretical explanation of stress argues that a person's physiological response results from his or her cognitions (Lazarus and Folkman, 1984).

However, other psychologists adopt a different position. To these psychologists, human beings are biologically primed to detect threats in their environment without the involvement of cognition. When presented with a challenging situation, our brain immediately registers the threat and precipitates the series of physiological responses that elevates our heart rate, our breathing, and so on.

Only after we feel this surge of physiological changes do we begin to think through what is happening. We proceed to evaluate our circumstances and, in turn, to interpret our physiological response at a cognitive level. If we feel the challenge is not very stressful, we might consider our increased heart rate and giddy breathing to be a sign of excitement. But if we feel the challenge *is* very stressful, we are liable to interpret our physiological reactions as manifestations of stress. This is plausible because, physiologically, stress is quite similar to excitement. In this theory, the psychological feeling of being under stress is a consequence of the physiological stress experience, not its cause (Schachter and Singer, 1962).

As theoretical models go, the symmetry here is striking. One explanation is effectively the opposite of the other. In the first theory, cognitions precede and precipitate our physiological responses; in the second, physiological responses precede and precipitate our cognitions. When two theories provide mirror-image accounts of the same psychological experience, it forces at least one firm conclusion: both cannot be correct.

So which is true? Well, we just don't know. Plenty of psychologists take the first theory – that cognitions drive stress responses – as their fundamental framework. From this they conclude that one good way to minimize or even prevent stress is to get people to think in different ways. But plenty of other psychologists take the second model – that stress responses drive cognitions – as *their* framework. They conclude that changing cognitions can be expected to have only minor effects on stress responses. So not only do the two theories stand in contrast with each other, they suggest directly contrasting courses of action. If you follow one theory you recommend one action, if you follow the second theory you recommend another. Not to put too fine a point on it, the theories are contradictory.

Both are considered mainstream psychological explanations of the human stress response; neither is considered controversial or contentious. Both are cited extensively in health psychology textbooks and both underlie hundreds of new research studies conducted every year. But if you believe one of these theories, it is intellectually difficult to believe the other at the same time. Technically, you should disbelieve it. Of course, some psychologists see their way to arguing that both theories *can* be true at the same time, but generally this requires something of a fudge (such as declaring the first theory to be true under some circumstances and the latter true under different circumstances). The fact that such expedience involves the development of a third, hybrid, theory is possibly one of the reasons there are so many theories in psychology.

Psychology's fragmentation

At the time of writing, the relevant Wikipedia entry lists an impressive *forty-five* different schools of psychology ('List of psychological schools', 2018). Having said that, it is generally agreed that psychology comprises around six major paradigms.

- The first is **biological psychology**. According to this theoretical view, psychology is best approached by drawing on principles of biology. This can often mean referencing physiology or neuroanatomy when building an account of human thoughts, feelings, and actions. It can also involve using other biological specialisms, such as genetics, pharmacology, or evolutionary theory. A biological psychologist might consider depression to be the consequence of unusually pronounced functioning of particular aspects of brain neurochemistry, which may in turn reflect some kind of genetic predisposition. In short, the assumption of biological psychology is that people's thoughts, feelings, and actions reflect the biological nature of human beings and their evolution as an organic species. This is the paradigm of *humans-as-animals*.
- The second major paradigm is **behaviourism**. As mentioned above, behaviourism focuses on the ways in which humans are influenced by events, or stimuli, in their environment. To a behaviourist, people's thoughts, feelings, and actions result from naturally occurring reflexes, which over

a lifetime are shaped by rewards, punishments, and other experiences. In behaviourism, depression may result when a person finds that adverse outcomes arise no matter what actions they take. In this context depression reflects a feeling of helplessness that is shaped by events in the environment. A major tenet of behaviourism is that behaviours are irreducible. This means that the study of behaviours should be the core activity of research and that it is not useful to look beyond observable behaviours in order to speculate about unobservable mental processes. In short, the assumption of behaviourism is that people's thoughts, feelings, and actions reflect their exposure to environmental stimuli and reinforcement. It is the paradigm of *humans-as-responders*.

- The third major school of psychology is **cognitivism**. Cognitivism places mental processes at the centre of psychology. It argues that any explanation of human psychology must account for what is going on in people's minds. It is often claimed that cognitivism emerged in response to behaviourism. However, scholarly reflection on the nature of mental processes has a very long history in philosophy. Cognitive psychologists study information processes such as memory, perception, and attention. Cognitive approaches to depression refer to the counterproductive nature of distorted thought patterns, such as the habit of focusing on negative aspects of situations instead of positive ones. In short, the assumption of cognitivism is that people's thoughts, feelings, and actions reflect the ways in which human minds process information and knowledge. Cognitivism is the paradigm of *humans-as-thinkers*.

- The fourth major paradigm is **social psychology**. Social psychology argues that human beings are inherently social creatures, whose patterns of thoughts, feelings, and actions can only properly be understood by looking at how they relate to the influence of other people. While social psychologists often study the behaviour of groups, they also consider solitary behaviour in terms of its social determinants and functions. To a social psychologist, depression is likely to result from deficiencies in an individual's social circumstances. It may be caused by poor quality relationships, family dysfunction, or social isolation. By default, social psychology emphasizes the influence of nurture over that of nature. In addition, by focusing on social contexts, such as communities and societies, it places a strong emphasis on how culture shapes the human perception of reality. For this reason, some social psychologists criticize the ways in which the rest of psychology attempts to describe (and measure) such humanly perceived reality. For example, some argue that the concept 'mental health' is a social or cultural construction, rather than something objective that truly exists in nature. In short, the assumption of social psychology is that people's thoughts, feelings, and actions reflect the impact of interpersonal relationships, group dynamics, and culture. This is the paradigm of *humans-as-members-of-groups*.

- The fifth major school of psychology is **psychoanalysis** or, as it is often referred to, the psychodynamic school. Psychoanalysis was first developed by Freud but has been elaborated upon by many others during the past century. It emphasizes the role of unconscious processes in human behaviour. Psychoanalysis argues that people's thoughts, feelings, and actions are largely determined by various instincts and impulses competing for dominance in their unconscious minds. One example is how a person's natural survival instinct competes with their instinct for self-destruction or self-sabotage. According to psychoanalysis, the exact balance between these competing impulses will be affected by the person's early-life experiences and relationships. To a psychoanalyst, depression might reflect a displaced feeling of grief that a person experiences as a result of a problematic early-life relationship, such as feeling rejected by a parent. In short, the assumption of psychoanalysis is that people's thoughts, feelings, and actions reflect the influence of deep-seated unconscious drives and instincts, the precise manifestations of which are determined by early-life events. Psychoanalysis is the paradigm of *humans-as-driven-by-instinct*.

- The final major school of psychology is **humanistic psychology**, a broad area that overlaps with philosophical movements such as phenomenology and existentialism. Humanistic psychology attempts to focus on aspects of human nature that are distinctively human, such as dignity, self-worth, and meaning. While offering some general models of the human experience – such as the theory that people are motivated to pursue a hierarchically ordered set of needs – humanistic psychologists tend to emphasize the individual uniqueness of each person. From the perspective of humanistic psychology, depression may result from particular circumstances that prevent a person from fulfilling his or her creative potential, such as a stifling marriage or job. Humanistic psychologists champion 'holism'. They argue that the scientific approach of developing generalizable explanatory principles is too blunt to properly account for human thoughts, feelings, and actions. For this reason, humanistic psychology is less associated with traditional research methods. In short, the assumption of humanistic psychology is that people's thoughts, feelings, and actions are a function of individual person-level perceptions of the world and the scope available for each person to strive for and achieve personal fulfilment and self-actualization. This is the paradigm of *humans-as-intrinsically-human*.

The patterned nature of incoherence

Variety, according to the cliché, is the spice of life. In that sense, then, at least we can say that psychology scores high on the Scoville scale. That might be as good as it gets. The fact that psychology contains so many paradigms could be its most serious flaw. It may be a sign that psychology has yet to become a mature science: philosophers have often posited that *real* sciences become homogeneous and internally coherent over time, as strong theoretical and methodological approaches prosper while weak ones fail.

Some psychologists suggest that this filtering has already taken place, with their own field emerging as the archetype. Behaviourists often argue that when their paradigm became dominant it displaced psychoanalysis. Cognitivists sometimes claim to have toppled behaviourism. Some modern biological psychologists – for example, neuroscientists – believe their own work to be so influential as to represent the future of psychology. Humanistic psychologists take the view that their field is meaningful in ways that mainstream science cannot ever be; thereby positioning humanistic psychology as the subfield that truly trumps all others.

But in the real world, becoming a dominant paradigm is not just a matter of producing the most reliable findings. After all, as we have seen, if research reliability is the key performance indicator, no paradigm in psychology will be found to have met its targets. In reality, the shape of a science is determined by rather more pragmatic factors. These include reputation and cachet. Scientists gravitate towards paradigms that are fashionable (or well-funded) as much as towards those that are empirically rigorous. They jump on bandwagons, ride on gravy-trains, and follow crowds. That is not to say that science has no sense of direction; after all, few paradigms become fashionable if they are truly useless. But in psychology's case, all the major theoretical schools seem to have demonstrable utility, in that all have attracted – and continue to attract – followers. None dominates. This as much as any other reason accounts for the heterogeneity of the field.

Perhaps the best example of endurance is psychoanalysis. Psychoanalysis is dismissed as scientifically unsound by nearly everyone, bar psychoanalysts themselves. Both philosopher Karl Popper (1965) and science commentator Lewis Wolpert (1992) famously chose psychoanalysis as their favourite example of nonsense. Nobel-winning physicist Richard Feynman (1998) considered psychoanalysts no more than 'witch-doctors'. Critics of psychoanalysis argue that its descriptions of intrapsychic machinations are too speculative, uncorroborated by firm evidence, and overly reliant on unprovable suppositions.

So intense has the criticism been that psychoanalysis is frequently depicted as a defunct or, at best, anachronistic branch of psychology. And yet, both as a set of theories about human psychology and as a basis for psychotherapeutic intervention, psychoanalysis remains extremely popular. Of course, popularity does not prove potency; it could be that psychoanalysis is perceived as helpful because it is so well liked, rather than the other way around. Nonetheless, voluminous droves of clients say that psychoanalytic psychotherapy is extremely helpful to them, and there are more members in the American Psychological Association's division for psychoanalysis than there are in its division for counselling psychology (APA, 2017).

Something for everybody

One consequence of paradigmatic heterogeneity is the coexistence within the mainstream of wildly contradictory perspectives, all of which enjoy the status of psychology's received wisdom. Scrutiny of any major introductory psychology textbook will reveal that the various schools are presented with more or less equal prominence. Students are required to internalize the lot.

Cognitivism teaches them that cognitions are the key driver of human thoughts, feelings, and actions; psychoanalysis teaches them that the key driver is the unconscious; while behaviourism teaches them that the unconscious doesn't exist and that cognitions are, at best, a mere side-effect of existence. Biological psychology teaches them that personality is predominantly genetic and heritable; social psychology teaches them that personality is predominantly shaped by contact with social role-models. Biological psychology teaches them that stress responses precipitate cognitions; cognitivism teaches them that cognitions precipitate stress responses.

Behaviourism teaches them that stress is best dealt with by accepting unpleasant thoughts and learning not to overreact to them; cognitivism teaches them that stress is best dealt with by extinguishing problematic thoughts and learning to think differently. Behaviourism, cognitivism, and biological psychology assert that using reliable measures of objectively verifiable observations is the best way to produce generalizable knowledge about the human experience; humanistic psychology asserts that the human experience is not generalizable.

By and large, psychologists seem extremely tolerant of such contradictions. There are a number of reasons for this. Firstly, some psychologists consider psychology's variety of schools to be a sign of healthy diversity. Their existence provides psychologists with *choice* and so fosters competition. Ordinarily, if people choose their favourite paradigm on the basis of quality, then it certainly follows that competition should *improve* quality: more productive approaches should be seen as more attractive, and less productive ones should fall by the wayside. The problem, as explained above, is that psychologists do not typically choose paradigms on the basis of quality alone (if they ever do so at all).

Secondly, some psychologists accept the widespread contradictions because they believe in freedom of expression (specifically, academic freedom). They consider psychology's different schools to be akin to society's different religions: it is desirable to accommodate all points of view in order to promote harmony. However, psychological paradigms are not like religions in that psychology as a whole is supposed to be considering a single truth. Religions are designed to offer competing realities. When society treats multiple religions with equal respect, it is agreeing that all are legitimate, not claiming all are true. Psychology, as a science, is supposed to produce knowledge that people can rely upon. This makes it somewhat problematic for psychology to endorse a multiplicity of mutually incompatible knowledge claims.

Finally, some psychologists accept the contradictions because they consider psychology's various paradigms to be complementary rather than conflicting. In other words, they don't pay the issue too much attention. While the paradigms might contradict each other on paper, these psychologists would like to avoid contradicting each other in person.

Narratives vs. numbers: How qualitative is more but quantitative is better

Not only does psychology lack a single view of human nature, it also lacks a single view of psychology. In other words, psychology is methodologically as well as theoretically sectarian. Some psychologists feel that their preferred methods are not just desirable, they are in fact the only approaches that should ever be used. They see alternatives as not just weak but irrational. They are unwilling to accept the merit of the other point of view.

The clearest example of this divide is that between quantitative and qualitative research. Quantitative research relies on traditional scientific approaches such as numerical measurement, experimentation, the statistical assessment of hypotheses, and the detection of generalizable principles that can be used to describe most people. Qualitative research focuses on the idea that the most important source of psychological insight is personal experience, both of research participants and of researchers. As such, qualitative research will involve approaches such as interviewing, observation, ethnography, and discourse analysis.

Quantitative research focuses on that which can be measured and independently verified. As such it will often limit its scope in order to ensure that its elements are replicable. For example, rather than asking participants to describe in general how they are feeling, they will ask them to complete standardized rating scales in which they can provide numerical scores that best represent specified emotions (e.g., a response to the question 'On a scale of 1 to 5, how happy are you right now?'). In this way, quantitative research is often described as involving reductionism (effectively, the attempt to explain complex phenomena in terms of simpler ones). But its procedures can be reproduced and its findings clearly replicated (or not, as the case may be).

On the other hand, qualitative research invites participants to describe their subjective experiences in their own words and at length. In this way it is said to involve a richer and more complex type of data. At a practical level, qualitative researchers often describe their studies as being much more labour-intensive than the work conducted by their quantitative peers. They don't like people to think they have it easy.

Technically, quantitative research is the default in psychology and other sciences. In contrast, qualitative research is an occasional practice, imported from related fields such as anthropology and philosophy. Indeed, the expression 'quantitative research' exists only because of the emergence of research of the qualitative kind. Note that the Wikipedia entry for 'Quantitative Research' uses the definition drawn from an encyclopaedia of *qualitative* research methods (Given, 2008). The term is never used by real scientists to discuss ordinary science.

Virtually every sympathetic account of qualitative research describes it as a response to, and an advance on, quantitative research. Typically, it is described as having overcome the problems associated with positivism. Positivism is the traditional scientific view that there is a straightforward relationship between the real world and our perceptions of it. According to qualitative researchers, a positivist (or quantitative) approach to psychology ignores the subjective manner in which people construct their own realities. It also ignores the fact that researchers themselves will have subjective views of their research questions, and indeed of their data. They argue that scientific objectivity is very difficult (if not impossible) in psychology and that attempting to conduct psychology research as if it were equivalent to research in, say, physics or chemistry is effectively futile.

In short, the qualitative critique of mainstream psychology is that it fails in its aim to be objective. In other words, it is fatally contaminated by subjectivity. However, the response to this problem is somewhat paradoxical. Qualitative researchers' solution to psychology's incorrigible subjectivity problem is to employ a set of methods that are themselves highly subjective. Subjectivity is seen as debasing psychology through quantitative research but rescuing psychology through qualitative research.

Qualitative researchers explicitly promote reliance on subjective data, the kind gathered through observation and interviews. They also promote the infusion of researcher subjectivity into the heart of the research process. Researchers are invited to reflect on data and, through their own judgements, to draw intuitive conclusions as to what is revealed. For example, the qualitative approach of 'grounded theory' requires researchers to study whatever information they have gathered (e.g., interview transcripts) until they are able to discern repeated ideas; these ideas are then grouped into categories and themes; and from these categories and themes, the researcher builds a theory.

This is said to be an improvement on the traditional scientific approach of presenting a theory in advance of research and then establishing whether or not the theory holds water by examining relevant data. A traditional scientist might argue that the grounded theory approach is overly reliant on researchers' subjective opinions. They might also complain that it creates conditions ideal for an epidemic of confirmation bias. The method makes a virtue of inviting the researchers to wait until the data are in before announcing their predictions. This is the epitome of what we described in Chapter 1 as 'HARKing' (or 'hypothesizing after the results are known'), a logical faux pas that tends to make non-replicable research inevitable.

Such make-it-up-as-you-go-along approaches are intuitively appealing. After all, it is what most people do in everyday life. Most human beings accumulate their knowledge about the world in a largely passive way, through incidental exposure to media, society, and other people. We occasionally engage in deliberate research; for example, we shop around for car insurance. But in most cases, we approach the task of finding things out in a relatively casual manner. We look things up on the internet, we ask people nearby what they think or know, and we cobble together our own recollections in a haphazard fashion.

Rarely do we attempt to systematically falsify our own assumptions. In fact, we usually do the opposite. As human beings, we are pre-programmed to bias our observations towards our own prejudices, and we focus on knowledge that reinforces our prior beliefs more than on information that challenges them. It is for this reason that philosophers first developed the idea of a scientific method. Natural human intuitive attempts at logic are usually flawed. We draw better conclusions when we step away from our instincts and outsource our reasoning to some systematic and independently verifiable method.

Scientific approaches represent a collective effort to bypass the frailties of subjective human judgement. Researchers who wholeheartedly embrace subjectivity would appear to believe such frailties are not worth bothering about.

Quantitative approaches are descended from positivist traditions of human knowledge and embrace a belief in empirical realism. In other words, they refer to the view that there exists a real world that can be observed by many people. Qualitative approaches are descended from a very different philosophical thread. They reflect traditions such as relativism and constructivism, theoretical views that reject the notion of empirical realism. In other words, they refer to the view that there is no empirical reality per se, only the observations and subjective perceptions of individual persons.

In short, quantitative and qualitative approaches differ in their assumptions regarding empirically verifiable reality. Quantitative approaches assume that empirically verifiable entities (a) exist, (b) constitute a superior form of knowledge, (c) trump anecdotal alternatives, and (d) represent intellectual rigour. Qualitative approaches, on the other hand, assume that empirically verifiable entities (a) don't exist, (b) constitute a delusion of knowledge, (c) are worthless compared to anecdotal alternatives, and (d) represent intellectual folly. As such, the two approaches present directly competing worldviews. If you consider one worldview correct, you must consider the other to be mistaken.

Why can't we all just get along?

In reality, of course, psychology is open to all-comers. When it comes to methods, it is customary for psychology to advocate the merits of everything at the same time, regardless of contradictions. Psychology textbooks describe both quantitative and qualitative methods dispassionately. Professional bodies, such as the British Psychological Society (2013), require that both be taught. As with any sectarian divide, researchers who build bridges are praised. In the research literature there are instances of peaceful coexistence, and even marriages (albeit arranged ones), between the two sides.

Some psychologists go so far as to promote the use of so-called mixed methods (Johnson, Onwuegbuzie, and Turner, 2007), in which researchers are called upon to employ *both* quantitative *and* qualitative approaches in their work. In other words, they are called upon to *both* accept *and* reject the notion of empirically verifiable reality, to *both* valorise *and* demean scientific objectivity, and to *both* avoid *and* embrace subjectivity in the research process.

Psychologists who adopt such liberal forms of agnosticism can become very critical of those who insist on separation. Quantitative (i.e., mainstream) researchers who dismiss qualitative methods are usually frowned at. Qualitative researchers who make disparaging remarks about the 'natural sciences' can expect to be rebuked. 'Mixed methods' advocates, meanwhile, express lofty fatigue whenever the qualitative-quantitative debate is aired, as though the distinction between contradictory research approaches is no longer a relevant concern to erudite people. It is as if they wish to prioritize harmony among psychologists above replicability of research.

But fundamentally, whether psychologists focus on the issue or not, quantitative and qualitative methods are very, very different. They differ not just in technicalities but in philosophies. They are based on competing assumptions about how human beings perceive their lives, what effective research should look like, and what reality itself is made of. Accepting both approaches as valid requires a level of intellectual detachment that might not serve psychology well in the long run.

Mixing contradictory methods is easy if you ignore the rudiments of the methods that are to be mixed. Combining contradictory assumptions will not matter if all you do is blindly apply a set of research procedures and disregard their underlying merit or rigour. The simultaneous promotion of both quantitative and qualitative research in psychology invites a superficial attitude to the research process, an eclecticism that promotes a colour-by-numbers approach to science. When psychologists are encouraged not to challenge the reliability of common research methods but, rather, to accept all as valid on their own terms, then it would be unwise to expect any resulting body of findings to display consistency, reliability, or replicability.

Garbage in, garbage out

Seeking to accommodate both quantitative and qualitative methods in psychology might sound academically open-minded. Showcasing psychology's various competing theoretical paradigms may appear liberal and healthy. A psychology that embraces traditional science as well as postmodern relativism is certainly diverse. One that promotes behaviourism alongside psychoanalysis is extremely so. However, such inclusivity does not guarantee reliability. In fact, it militates against it.

When asked for the causes of depression, psychologists are liable to answer that depression is caused by a chemical imbalance, reinforcement of despair, cognitive distortions, lack of social support, displaced grief, and/or an inability to fulfil one's creative potential. Professional bodies and textbooks will assert that all such answers are reasonable. Asking one psychologist will yield one response. Asking another will yield a different explanation. Empirical research is not needed in order to produce a replication crisis in psychology. Psychologists can engage in scientific disputes all by themselves, without the requirement for conflicting research findings.

Inconsistency in the starting positions from which scientific research is embarked upon will guarantee inconsistency in the findings that emerge from such research. Psychology's propensity to afford equal airtime to contradictory worldviews (both theoretical and methodological) not only bolsters its own form of a post-truth culture, it escalates the risk of poor replicability. It feeds psychology's replication crisis.

CHAPTER 3

'Never Mind the Quality, Feel the Width': Psychology's Measurement Crisis

Intelligence, real and artificial

Everybody loves a robot. Whether a gold-coloured mannequin with perfectly circular eyes or a stumpy barrel-shaped contraption with three wheels, the fact that it can think and communicate like a person excites our imaginations. Humans are intrigued when humanlike intelligence is exhibited by non-human things. It is not so much the technological achievement that enthrals: it is the sheer *mystical* quality of a 'thinking' machine, an artificial device endowed with psychological power.

Warriors made of gold and clay have been mentioned in myths for centuries, and androids have appeared in cinema since the days of the first silent movies. Robots and androids are long-standing staples in literature and drama. In fact, the words 'robot' and 'android' were both coined by writers of mass-market fiction: the former in Karel Čapek's 1920 play *RUR: Rossum's Universal Robots* (Čapek & Selver, 1923), and the latter in *The Future Eve*, a novel by Auguste Villiers de l'Isle Adam (1886).

More recently, artificial intelligence has become the target of extensive research effort. Machines can now mimic many human cognitions and learn sophisticated problem-solving techniques. One example is the development of self-driving cars, which can navigate and decision-make more efficiently than any human driver. Not only can the car change lanes by itself, it can even signal properly when negotiating a roundabout (a task most humans seem utterly unable to perform). Driverless cars are not just more capable of obeying the highway code. They also obviate risks associated with drunk-driving, phone calls at the wheel, road rage, and carpool karaoke. Smart cars ensure safe people.

So impressive is the march of artificial intelligence tech, some futurologists now speculate about the 'singularity' (Vinge, 1993), the day when machines become so smart, they reach a point of superintelligence. The smart machines start producing smarter machines all by themselves, each in turn going on to produce even smarter ones, prompting an abrupt surge in the power of machine-based braininess.

According to some commentators, the singularity will usher in an amoral technocratic era characterized by mass unemployment, robot wars, and eventual human extinction. For others, however, the singularity is akin to utopia. It will provide the processing power to enable decades' worth of research to be completed almost instantaneously, thereby allowing humanity to solve all its medical, environmental, and societal challenges. According to Masayoshi Son, the CEO of one of the world's largest telecommunications corporations, the singularity is likely to occur within the next three decades (Shead, 2017). By then, he claims, the average computer will be so smart, it will have an extremely high IQ. To be precise, he says that it will boast an IQ score of 10,000 (Son, 2017).

Most psychologists will tell you that intelligence is hard to measure. It is no small challenge to attach a score to a cognitive ability. Psychological concepts are by their nature amorphous, and so their measurement represents one of the greatest challenges facing behavioural science. In fact, as we will see in this chapter, psychological measurement is far from the exact science that non-psychologists assume it to be. But this CEO is an experienced technology expert, accustomed to mingling with high-end computational geniuses. He must know what he is talking about. His predicted IQ of supercomputers – 10,000 – seems confidently specific. So, then, how does it all stack up?

No match for natural stupidity

Sometimes it seems that talk of artificial intelligence leads people to say things that are, well, not that smart. Take, for example, this idea of a superintelligent computer's IQ. IQs are not quantities like weight or length that can be measured using scales or callipers; rather, they are scores achieved on assessment tests devised by psychologists. So what is being said is that if you tested a post-singularity computer's intelligence, you would see it get an IQ score of 10,000.

mental age
bio age (handwritten marginal note)

At a very basic level, IQ stands for 'intelligence quotient', indicating that it represents a ratio (or 'quotient') between two values. When measuring intelligence, these values are a person's biological age and the age that best encapsulates their mental abilities. You compute IQ by dividing the former into the latter. A ten-year-old who thinks like a twelve-year-old will have an IQ of 120 (namely, twelve divided by ten; for clarity, IQ scores are expressed in terms of hundreds). A ten-year-old who thinks like an eight-year-old will have an IQ of 80. And a ten-year-old who thinks the way most ten-year-olds think will have an IQ of 100, making that score the natural (albeit contrived) benchmark of 'average' intelligence.

It is useful to keep in mind that IQ tests were first developed for children. The system makes sense in children, because a child's intelligence increases annually with age, in line with their brain development. But in adulthood intelligence does not grow year by year. Adults do not become proportionally cleverer with each passing birthday. A twenty-year-old might be really smart, but it does not make sense to say they have the brains of a thirty-year-old. When Albert Einstein was sixty-five years old, he was said to have an IQ of 160; but this was not meant to imply that the man's mental abilities resembled those of a centenarian.

Adult IQs should not be interpreted literally as adult-age differences but as analogues of the relativities seen in children. Where adults with IQs of 120 are smarter than adults with IQs of 100, the gap is intended to be equivalent to that seen between twelve-year-olds and ten-year-olds. Similarly, Einstein's IQ was 160 because his superiority to ordinary adults reflected how children aged sixteen outshine children aged ten.

An important point to note is that measurements of IQ do not represent *amounts of* intelligence. In other words, Einstein's IQ of 160 did not mean he was 1.6 times smarter than the average person. Instead, IQ represents a *symbolic relativity between* intelligences, with the range seen in children used as a frame of reference. To say that a computer (or a person) might one day have an IQ of 10,000 stretches what could make sense. Literally, an IQ of 10,000 refers to the difference you might see between a child of a given age and another who is 100 times older.

The obvious problem is that childhood brain development does not easily facilitate such comparisons, at least in terms of what IQ tests measure. In order to exhibit verbal intelligence, for example, a child needs to be able to speak. The average age at which children start talking is one year. Therefore, an IQ of 10,000 should be benchmarked on the comparison between a one-year-old child and a 100-year-old child, which of course is nonsensical, even as a basis for hypothetical extrapolation. It is partly for this reason that most IQ tests are designed (i.e., rigged) to produce a more-or-less sliding scale of scores with a maximum just above 160. This approximates the relativity in age between a verbal one-year-old baby and a child aged sixteen (the age commonly seen as the end of childhood).

The fact that IQ scores represent *relativities* rather than *amounts* is rooted in biology. The measurements produced by IQ tests are not quantitative dimensions. They are after-the-fact labels attributed to people's abilities based on how far their test scores fall above or below the average. IQs are produced by tests designed with the express intention of rating the capacities of living human beings in comparison to one another. In order to be useful, they are deliberately set up to produce scores that are, in statistical language, 'normally distributed'. This means that scores spread in such a way as to disproportionately load around the notional midpoint of 100, with high and low scores arranged against standard thresholds. If your performance is among the top 5 per cent of people, you will be awarded an IQ above 120; if it is in the top 2 per cent, your IQ will be above 130; if it is in the top 0.1 per cent, your IQ will be at least 145. If you achieve the maximum score, by answering all the questions correctly within the time allowed, your IQ will be around 160 (the exact score depends on which test you use; Hunt, 2011). If this happens, you will be in an extremely rarefied group. Even among the members of Mensa, the society for people with extremely high IQs, only 1 per cent of people score 160 or more on their IQ tests (Gani, 2015).

Aside from this, to claim a computer can have an IQ of 10,000 is problematic for other reasons. Chief among them is the fact that intelligence is not what we call a 'unitary construct'. This is a complicated way of saying that it comes in many shapes and sizes. One person can be smart at mathematics, whereas another can excel in verbal reasoning. A third might be precocious at detecting patterns in complicated information. Another might be exceptionally quick at comprehending new ideas. Somebody else might be gifted at understanding how people think. Another person might have a unique talent for music, able to reproduce a complex tune after hearing it just once. Simply put, there is more than one way to be smart.

A common criticism of IQ tests is that they fail to account for the full range of intelligences. That said, most of them will at least attempt to measure different abilities, and underlying any IQ score will be a set of subscores from the various sections that comprise the overall test (such as vocabulary understanding, arithmetic ability, memory for sequences of numbers, ability to detect what is missing from a complex picture, and so on). Many such tasks are mechanical and so can be performed easily by even very elementary computers today. But a machine will always find it difficult to compose a memorable sonata, to produce new insights about the origins of the universe, or to choose a gift that truly excites a lover's passion. While it is reasonable to suggest that computers can store lots of data and process information very quickly, to claim they possess de facto intelligence seems unsound.

IQ is a concept invented by humans to refer to measurements of intelligence in other humans. Speculating about an IQ over 160 is flimsy, even as a metaphor. Go as far as 10,000, and the analogy makes little or no sense. When a tech CEO predicts that computers will one day boast 10K IQs, it is unclear (a) how such a score might ever be computed, (b) what such a number signifies, and (c) what type of intelligence(s) such a figure might actually refer to. The flaw in the claim can be exposed by the following question: if computer IQs will one day reach 10,000, then what IQs do *contemporary* computers possess? What, for example, is the IQ of my laptop?

The quagmire of psychological measurement

If it is to be useful, psychological measurement requires both accuracy and authenticity. The outcome needs to be precise, but it also needs to truly encapsulate the actual thing that is being measured. In psychology, this is much easier said than done. For example, intelligence tests need to produce exact scores, but they also need to faithfully represent whatever it is we mean by the term 'intelligence'. Both requirements are difficult to meet. Providing exact scores is difficult because intelligence is multidimensional, fluid, and subtle. Faithfully representing what we mean by 'intelligence' is difficult because, in effect, psychologists cannot agree on what is meant by the term. Measurement is not just a procedural or statistical task; it is conceptual and theoretical too.

Measuring things is by far the trickiest part of conducting psychology research. It is a greater challenge than designing decent experiments, recruiting representative participants, or analysing statistical data. But in some ways it is a hidden challenge. It is relatively easy to measure things badly but relatively difficult to know when you are doing so. Moreover, bad measurement is fatal to good research. Even the most sophisticated study will be pointless if the variables of interest are poorly measured.

In psychology research, everything starts and ends with measurement. The quality – and replicability – of studies depends on how seriously researchers take the task. Measurement is profound: it is the moment when psychologists get to say 'There it is!' But, as with many method-ological subtleties, thoughts of rigour occupy the minds of only a minor-ity. Some psychologists obsess over the job of getting measurement just right; others blithely ignore it.

It ain't (only) what you do, it's the way that you do it

The accuracy and authenticity required by good measurement are more technically referred to as 'reliability' and 'validity'. In short, measurement *reliability* is the extent to which a method produces consistent scores. If people get the same scores each time they do a particular IQ test, we say the test has good reliability; if their scores vary, we say the test has poor reliability. Measurement *validity* is the extent to which a method succeeds in measuring what it is supposed to measure. If an IQ test has high valid-ity, then people of genuinely high intelligence will receive high scores, people of average intelligence will receive average scores, and so on. If it has low validity, then there will be little or no association between people's actual intelligence and the scores they receive on the test.

There are many ways in which psychological measurement can end up with poor validity. The technique used to measure a concept might be flawed. Alternatively, the concept itself might be so poorly defined as to defy attempts to measure it. Or the meaning of a concept can be changeable, ambiguous, or opaque, all of which also undermine measurement validity.

Take, for example, a notion such as internet addiction. It is common to encounter anxiety about the amount of time children, teenagers, and adults spend on the internet. Excessive internet use is claimed to lead to serious mental health problems, and concerned media columnists regularly call for the curtailment of children's access to smartphones, tablets, and computers. This sounds very much like something psychologists could usefully investigate. Nearly seventy large-scale population studies have been conducted into internet addiction to date (Kuss et al., 2014). However, despite the investment of time and resources represented by this research effort, the findings produced so far seem truly muddled. The biggest problem is poor validity.

One sign of this is the dramatic divergence in rates of internet addiction seen across different studies. A study in Hong Kong found that 26.7 per cent of young people suffered from internet addiction (Shek & Yu, 2012), whereas a study in Italy found that only 0.8 per cent were so afflicted (Poli & Agrimi, 2012). While it is possible for pathologies to arise with different frequencies in different places, such variation in prevalence seems quite extraordinary. Revealingly, in order to measure internet addiction, the studies used two different versions of the same questionnaire, one written in Chinese and the other in Italian. This immediately raises the suspicion that perhaps the notion of internet addiction got lost in translation. Maybe the findings diverged because the researchers ended up measuring different things. In other words, one group of researchers measured internet addiction as if it were one thing, whereas the other group measured it as if it were something else. If so, it can hardly be said that both groups measured it correctly. One of them, or both, got it wrong.

Research into internet addiction has been plentiful, but useful findings have been rare. A big problem is the changing nature of the internet itself. We might have hoped that psychologists could have kept up with developments a little more assiduously. The most popular clinical questionnaire for assessing internet addiction – and the one used by the above-mentioned Hong Kong and Italian studies – is the Internet Addiction Test, a test designed in 1998 (Young, 1998). The obvious problem is that in 1998, the internet was very different from what it is now.

For one thing, in 1998 the internet was a relatively little known service. Personal computers didn't come with preinstalled web browsers; users interested in this new 'internet' were required to install a browser all by themselves, using a CD procured from an internet service provider. Email was limited by download speed and inbox size. Google barely existed (its homepage was still marked 'BETA'). For most people the internet was a text-only experience. Online video was unheard of; it would be another seven years before YouTube was launched. The closest things to modern social media platforms were Bulletin Board Systems or similar services such as Usenet and Six Degrees. There was no Facebook, no Twitter, no tablets, no smartphones, and certainly no smartwatches. The idea that a person could (or would ever want to) be permanently connected to the internet was almost unconscionable.

The internet of 1998 was existentially, socially, and culturally a very different thing from what we see today. And yet, when you read about internet addiction in today's research, it is likely that the questionnaire used to measure it was developed with 1998 in mind.

One example is a widely reported brain imaging study claiming that addiction to the internet causes the same type of brain damage as addiction to narcotic drugs (Lin et al., 2012). While published in 2012, the study measured internet addiction as though it were still the late 1990s. The participants were seventeen years old, but the scale used to diagnose their addictions dated back to their infancy. The questionnaire was so dated, it hyphenated 'on-line' and spelled 'internet' with a capital 'I'. It asked them, 'How often do you check your email before something else that you need to do?' even though teenagers in 2012 would have thought it wholly uncool to use email for daily communication, no doubt preferring instead one of the many social media platforms that had by then become ubiquitous.

Several studies conducted in China have used a slightly different questionnaire, the Chen Internet Addiction Scale (Chen et al., 2003). However, as this scale was developed in 2003, it too is quite old, predating Facebook, Twitter, and YouTube, as well as all today's popular Chinese services, such as WeChat, Weibo, and QQ. In the end, while there are lots of large-scale epidemiological studies as well as some detailed brain-imaging experiments, the importance of this work is undermined by poor measurement validity. The findings produced are devalued to the point of being almost worthless, because a main component of the research – to measure the impact of internet addiction as it affects the population being studied – cannot be properly achieved.

Keep calm and carry on (regardless of measurement validity)

The shifting sands of cultural context lead to widespread problems with anachronistic measurement in psychology. Consider the example of life stress. One of the questionnaires most commonly used to assess exposure to life stress, the Social Readjustment Rating Scale (SRRS), was originally drafted in 1967 (Holmes & Rahe, 1967). The scale asks respondents to study a list of forty-three possible life events and to tick off all those they have faced in a given timeframe (for example, during the past year). Each life event has a weighted score representing its relative stressfulness. The authors developed these weightings by studying data gathered from large standardization samples.

They concluded that the most stressful event possible is the death of one's spouse, to which, for clarity, they gave a standard weighting of 100. Each of the other stressors was given a weighting of less than 100, which the authors carefully calibrated to reflect relative stressfulness. For example, 'Death of a close family member' was given a weighting of 63, while 'Being fired at work' was weighted as 47. The weightings were intended to facilitate the quantification of total life stress, and the original standardization data suggested that total stress correlates strongly with a person's risk of developing a physical illness. One interesting feature of the SRRS is its inclusion of several events that would generally be considered positive. For example, 'Marriage' has a weighting of 50, suggesting it is half as stressful as having one's spouse die.

Modern stress research often emphasizes personal resilience as the major determinant of stress-related well-being, and so scales such as the SRRS are less frequently employed than they used to be. However, the SRRS continues to be extensively cited and used in stress research. For example, one recent study linked 'experienced stress' as measured using the SRRS with age-related slowing of reaction time (Marshall, Cooper, & Geeraert, 2016). Another linked SRRS-measured life events with functioning of the immune system and physical aging of white blood cells (Lopizzo et al., 2017). In a study that used fMRI (functional magnetic resonance imaging) to examine brain anatomy, life stress as measured using the SRRS was found to predict the volume of the amygdala, the part of the brain associated with fear emotions (Sherman et al., 2016). Other researchers reported that SRRS-stress predicted depressive symptomology in patients who had a rare heart condition (Hintsa et al., 2016). In short, the SRRS continues to be used to measure stress exposure in a very diverse range of research contexts.

The problem, of course, is that the SRRS measures stress as it was experienced in the late 1960s, some five decades ago. In most societies, especially in Western culture, social attitudes have developed in ways that fundamentally alter these experiences. For example, according to the SRRS, the second most stressful event a person could ever experience in life is divorce. Divorce has a weighting of 73, making it substantially more stressful than even the death of a close family member (for example, a son or daughter). Interestingly, the SRRS lists 'marital separation' as a separate stressful event, with a lower weighting of 65, implying that the legal reality of divorce exerts significant additional impact. If a person was separated and then divorced in the same time period, their cumulative SRRS score for these events alone would be 138, much higher than if their spouse had simply died.

This almost certainly reflects a late 1960s societal attitude that was much more disapproving of divorce than is the case today. In the 1960s, divorce was generally considered to be irreligious and to run against the public interest. In most countries, including in the United States where the SRRS was developed, people could only get divorced if it were proven that one of the partners was guilty of violating the trust of the other (such as by cheating on them or abusing them). In other words, divorce was permissible only if one of the partners was 'at fault'. People who got divorced had to endure significant stigma and a risk of social exclusion. So-called no-fault divorces, in which people could divorce amicably, were not introduced to the United States until 1969 (and even then, only in the state of California). Divorce has since become far more socially acceptable. Its SRRS weighting of 73 seems grossly inflated by today's norms.

Many SRRS items relate, either directly or indirectly, to the current state of the economy. 'Changing to a different line of work', 'Being fired at work', 'Taking on a mortgage', 'Troubles with the boss', and so on, will all be more or less stressful depending on whether alternative employment is difficult or easy to come by. In the 1960s, the job market was considerably less fluid than it is today, and most people who entered employment were expected to keep the same job, or at least the same career, throughout their lives. In the modern economy, this traditional model has all but disappeared. People switch jobs very frequently and the idea of a 'job for life' is a lot less common. Thus, the context in which one might change one's employment today is fundamentally different from what prevailed in the late 1960s, and so the SRRS weightings for such life events are very likely now to be obsolete.

Finally, some of the SRRS items are anachronistic for slightly different reasons. For example, marriage is a very different life event today from it would have been in 1967. In the late 1960s fewer people lived together before getting married, and for many the prospect of matrimony was all the more daunting as a result. In 1967 participants who indicated that they had recently been married might as well have been ticking a box denoting they had just lost their virginity. Today marriage is undoubtedly a significant life event, but whether it is quite as existentially stressful as it was in the late 1960s is open to question.

Diagnoses (and second opinions)

Internet addiction and stress are examples of psychological concepts whose meanings shift naturally with the passing of time. However, sometimes meanings are shaped and reshaped in more deliberate ways. For example, official descriptions of mental health disorders are formally published in clinical directories, such the American Psychiatric Association's *Diagnostic and Statistical Manual of Mental Disorders* (DSM). These directories help clinicians choose correct diagnoses for clients who present with mental health problems. As new ideas and opinions emerge in fields such as psychiatry and mental health all the time, it has proven necessary to revise these diagnostic guidelines every decade or so. The problem is that rolling updates greatly undermine measurement validity in mental health research.

Consider the case of clinical anxiety. Up to 2013, a formal diagnosis of an anxiety disorder required a person to be aware that their anxiety was excessive relative to their situational demands. However, in the latest version of the DSM this prerequisite has been removed (American Psychiatric Association, 2013). Now you can be diagnosed with clinical anxiety even if you feel your anxiety is proportionate to your circumstances. In other words, you no longer need to be self-aware to have an anxiety disorder. That people can be oblivious yet still receive a diagnosis marks a fundamental shift in the concept of clinical anxiety. Therefore, it is unsound to consider anxiety research conducted before the revision to be comparable to research conducted after the revision. In effect, the newer research is examining a different experiential diagnosis in an expanded subgroup of the population.

Changes to the description of bulimia raise similar issues. A bulimia diagnosis used to require the presence of symptoms for six months or more. In the most recent DSM, this timeframe was reduced to three months or more. As with anxiety, many people diagnosed with bulimia today would have been declared healthy in the recent past. Therefore, modern research on bulimia will frequently include people who would have been excluded from earlier studies. In reality, the diagnostic criteria of many mental health disorders have been revised at various times. Therefore, it should be no surprise if studies conducted across different decades produce results that appear inconsistent.

Descriptions of mental health disorders are updated with the noblest of intentions. The aim is to enhance the lives of people who might benefit from more refined diagnoses. Some people will be guided towards mental health services they might otherwise have been denied, whereas others will be spared the burden of an unnecessary or inaccurate label. However, these updates serve to complicate the task of research. Questionnaires and scales for measuring anxiety, bulimia, and several other conditions whose criteria have been clinically reformulated, have seen their validity greatly diminished.

Construct validity: The root of psychology's measurement problems

In research, there are three main types of validity. The first is what we call _construct validity_. This is the degree to which something is what we say it is. It represents 'authenticity' in its purest sense. If construct validity is ropey, then psychology as a whole will be in trouble. There is little point proceeding with research if measurement approaches are flawed.

Construct validity is the extent to which our measurement method truly encapsulates the concept (or 'construct') we want to measure. Just because a psychologist gives a variable a specific name does not imbue its measurement with construct validity. If an IQ test is so poor that it measures something other than true intelligence, then simply calling it an 'IQ test' does nothing to resolve the problem. An internet addiction test is not a good measure of internet addiction just because it has 'internet addiction' in the title. Measures of life stress are not valuable simply because they are widely used. In psychology valid measurement requires more than just a literature search to find the most popularly used scale for a given variable or one with a convincing name. It requires a judicious assessment of the content of a scale, reflection on its fidelity to the variable of interest, and an assessment of its possible obsolescence.

When a concept is poorly defined, it is hard to measure accurately. When the meaning of that concept changes over time, it throws our measurement off. At best, we end up measuring a fuzzy version of what it is we want to study. We may even end up measuring something that isn't really there at all. We say 'There it is!' while pointing vaguely in the general direction of the wrong thing. This is what we mean when we say there is a difficulty with construct validity.

The big problem for psychology is that many of our key concepts are in fact abstractions. They signify entities that we assume exist, even though they are hidden from view. Thoughts, feelings, memories, opinions, attitudes, experiences, intuitions, views, ideas, moods, judgements, perceptions, preferences, and insights occur solely in the minds of beholders; a researcher cannot observe them directly.

Researchers who study a person's temperament or personality are restricted to doing so by observing them at a superficial level: they presume that people who go to parties are happy to do so; they use people's outer social activity as a basis on which to infer their inner personal nature and then classify them as extraverts. Researchers who study people's abilities or competences examine end-product rather than underlying process: they see somebody score high on an IQ test, have evidence only of their IQ test performance, and conclude that this reflects some deep-lying intelligence based on the assumption that it must do so. Researchers who examine social support gather the testimony of the person who receives it (or maybe the person who provides it) and make inferences about its supportiveness from hearsay because it cannot be quantified directly. Researchers who investigate the effectiveness of psychotherapy draw conclusions about client improvement based on symptom checklists or third-party diagnoses: it is not possible to conduct a blood test or brain scan to confirm the results.

The fundamental problem with indirect or inferential measurement is that it brings no guarantee of accuracy. To articulate and then measure an amorphous concept involves a process of abstraction that a researcher cannot account for in detail. It is unclear whether (or by how much) the construct is adulterated by the process of measurement or if the measurement is so indirect as to actually *mis*direct the measurer.

Participants' reports of even their *own* thoughts, feelings, and perceptions may simply be inaccurate, due to forgetfulness, distorted perspective, or the pressure of social desirability. People's visible behaviours may have essentially invisible causes (for example, they may attend parties for reasons other than extraversion). A high score on an aptitude test may be the result of a fluke or of an unanticipated or unfair advantage enjoyed by the test-taker (such as a test in which the questions relate closely to a person's own experiences or social setting). Descriptions of social support may be misleading because people often politely praise others for supportiveness, even when the help received is inconsequential. Reported improvements after therapy may result from optimism or false hope (on the part of clients and/or therapists) rather than from genuine recovery.

If we could avoid indirect measurement, we would. But the nature of psychology is that it studies the non-corporeal aspects of humanity. As such, it is always grappling with – and thus attempting to measure – the intangible. That said, we should note that even the measurement of tangible entities is far from straightforward.

Psychology is often contrasted with physics, where it is said that scientists can measure weight, distance, volume, and so on, without controversy. However, the assessment of physical properties, such as length or distance, will also involve inference. We measure physical dimensions by comparing objects to precalibrated length-defining tools (such as rulers or measuring tapes) or by recording the time taken for a signal (such as a sound or light wave) to travel the distance being assessed. Such methods appear so accurate as to be completely precise, and yet they are based on secondary approximation. The imprecision of these approaches is revealed when the distances being measured are large, such as when we attempt to send a space probe to a faraway planet, only to see it miss its target by thousands of miles. True, measuring a person's height is more straightforward than measuring their intelligence. However, this is because height is easier to define than intelligence. Once defined, the task of measuring either construct will still be indirect and inferential.

Construct validity is central to the success of science, and thus of psychology. However, in psychology it is extremely difficult to achieve. The nonspecific and fluid definitions of terms present ever-moving targets for researchers to aim at. In science, the pragmatics of research require scientists to offer clear working definitions of terms, so that studies can be independently replicated. One of the reasons psychology suffers from poor replicability is that its definitions are intrinsically vague rather than clear.

Internal validity: Measuring how things happen

The second main type of validity is *internal validity*. This is the extent to which we can use research to determine cause and effect. More loosely, it refers to our confidence in establishing the facts that have led to a particular situation. Any problem that leads a researcher to misinterpret causality undermines internal validity.

This is conspicuously relevant when psychologists study interventions. Simply put, intervention research requires psychologists to measure the functional status of participants *before* and *after* the intervention occurs. For example, psychotherapy researchers will want to measure how people are *before* therapy and whether their well-being is different *after* therapy. In an education study, researchers might test the abilities of children *before* a new pedagogical method is employed by their teacher and again *after* the method has been used. Research into awareness-raising campaigns will measure the public's awareness *before* the campaign has been run and again *afterwards*. The range of possible interventions is almost endless. Unsurprisingly, therefore, intervention studies are extremely common in psychology. In terms of public interest, they might even be the most valued type of study that psychologists are ever asked to conduct.

It is generally well known that simple before/after comparisons are insufficient to establish whether an intervention has worked. Just because people feel better after therapy does not mean that therapy caused their improvement. Just because children score better after a new teaching method does not mean that the method helped them. Just because the public are more aware after a campaign does not mean that this occurred because of the campaign.

Changes in outcome can happen for many reasons: maybe because of chance, maybe because of the passing of time, maybe because of the novelty of having witnessed an intervention (rather than having experienced its specifics), or maybe because of some other reason. Therefore, a proper intervention study will require, at the very least, a control group: namely, a second group of participants who do not receive the intervention. Scrutiny of change in this 'untreated' group can provide a benchmark against which fluctuations in the intervention group can be compared.

These days, such principles are widely understood. Even non-researchers will be suspicious of a study that lacks a control group (or control condition). But that is not to say that all studies with one will be perfect. While just about any credible researcher will know to include a control group, this does not guarantee that the control group so included will be sufficient. It is not a case of 'any control group will do'; some are better than others.

The commonest problem is when the circumstances of the control group are so different from those of the intervention group that comparisons are more or less pointless. When investigating whether people like a particular brand of coffee, it is more useful to compare it to another brand of coffee than to, say, tea. Weak control groups are arguably worse than no control group at all, because they create a false sense of security that breeds over-confidence in researchers, making them more likely to be lulled by false-positive findings.

Take for example a recent study of an intervention intended to improve people's attitudes to generic, as opposed to branded, medicines (Colgan et al., 2016). The authors studied a sample of seventy men and women who frequently suffered headaches. They presented half the participants with the intervention: a video featuring medical experts describing the benefits of generic medicines. The other participants, who made up the control group, were shown a different video: one that featured experts who talked about headaches but who did not mention generic (or any other) medicines. Afterwards, participants in the intervention group reported feeling more positive about generic medicines than did participants in the control group. The researchers duly lauded the intervention video for having been 'effective'.

However, all that was shown was that some participants repeat things they hear experts say in videos. It could be that the participants who were shown the intervention video simply wished to conform with the views presented to them. Perhaps social desirability encouraged them to mimic the experts' comments or to worry about losing face should they be seen to dissent from professional opinion. They may even have chosen to say what they felt the researchers *wanted* to hear (a common reaction that methodologists call an 'expectancy effect'). In other words, the video might not have changed their *attitudes* toward generic medicines. It may just have changed the *answers* they gave when faced with questions from researchers.

A good way to have excluded this possibility would have been to use a different type of control video. The logic of the study was confused by a poorly designed control group. If, like the intervention group, control participants were shown a video featuring experts talking about generic medicines – but this time, presenting *criticisms* of such medicines – then the researchers could perform a clearer comparison. They could better assess the degree to which participants were inclined simply to follow opinions that they saw on-screen.

In fact, the weak internal validity of the study was highlighted by another of its findings. When participants next had a headache, they were given either generic or branded medicines to take. Intervention group participants reported finding the generic medicines *less* effective, compared to control group participants. It appears that the video intervention had backfired, heightening participants' awareness of generic medicines so that they evaluated the pills more consciously and thoroughly than participants in the control group were inclined to do. Again, this would have been clarified had the control video corresponded better with the intervention video. In other words, the problem would have been lessened if the research had better internal validity.

The validity trade-off problem

Psychologists are sometimes accused of focusing too much research attention on *that which can be measured* and not enough on *that which is authentically of interest.* They spend their time looking at 'IQ test performance' instead of studying 'intelligence'. They employ checklists of life events that were stressful fifty years ago instead of methods that evaluate true emotional stress today. Construct validity is undermined by fuzzy definitions, but the problems are not clearly solved by making definitions less fuzzy. The issue is one of a trade-off between validities. Adding detail to a definition may help with measurement, but it can cloud the attribution of causality. The effort to augment construct validity can end up hurting internal validity.

Striving to specify the many details of a concept can lead psychological measurement to miss its target completely. Among other pitfalls, psychologists can mistake by-products for features. It is similar to the Indian folk tale where blind men study an elephant: one man touches the elephant's leg and thinks it is a pillar, another touches its tail and believes it to be a rope, another touches the trunk and mistakes it for the branch of a tree, and so on. By describing a concept in detail, psychologists can end up focusing on parts without properly taking account of the whole. And just as in the folk tale, where the blind men disagree with each other about what they have been examining, psychologists can produce diverging conclusions about the thing that has been measured, thereby creating on the conditions for non-replicability.

For example, clinical depression is a complex mood disorder characterized by many different symptoms. As well as negative emotional mood, these can include weight loss, sleep disturbance, loss of sex drive, loss of appetite, feelings of hopelessness, and so on. However, many of these symptoms are not just signs of depression. They can occur for other reasons too. Sudden or dramatic weight loss, for instance, can be a sign of several physical illnesses. Research complications arise when weight loss is considered so integral to the definition of depression that it is included in depression diagnosis questionnaires. Several major questionnaires do this. The Beck Depression Inventory (BDI; Beck et al., 1961), the Hamilton Depression Rating Scale (Hamilton, 1960), the Major Depression Inventory (MDI; Bech et al., 2001), and the Zung Self-Rating Depression Scale (Zung, 1965) all include questions about weight loss as part of their assessment of depression. If respondents tick the box to say they have recently lost weight, they will end up with a higher depression score than had they not lost weight.

The problem occurs when psychology researchers measure depression without accounting for possible physical illness. Ironically, sometimes the research question is specifically whether serious physical illness *causes* comorbid depression. If weight loss is the result of illness, then everyone who is ill will tick the weight-loss box in the depression scale and duly receive a higher depression score. This will happen even if, emotionally, a person feels no more depressed than normal. Such inflationary forces will give ill participants higher depression scores than well participants and could lead an imperceptive researcher to conclude that illness 'causes' depression.

The same problem will arise with respect to disruptions in appetite, sleep, energy levels, sex drive, and so on. The main depression questionnaires all include items asking about these symptoms too, and all attribute higher depression scores to respondents who report them. In reality, these symptoms could be the result of any number of physical illnesses. Some illnesses cause *all* these symptoms at once. Patients with such an illness will have super-high depression scores, even if they are in good spirits, optimistic about their lives, and in an authentically positive mood all the time.

This problem is called 'confounding'. It occurs when researchers are unable to tease apart variables in their studies. They mistakenly consider them to be overlapping when in fact they are separate, or perhaps they think they are separate when in fact they overlap. When psychologists define concepts in close detail, specifying all of their features, the risk of confounding is generally heightened. Some of those features might not be exclusive to the concept being defined. Treating physical symptoms as part of depression, as opposed to coincidental to it, can lead to false conclusions about a person who has these symptoms but who might simply be physically ill. In this case, we would say that depression and physical symptoms are confounded. Multifaceted definitions of depression – or of any other variable – are intended to boost construct validity. But in doing so they hamper our ability to determine cause and effect. In other words, they undermine internal validity.

When one becomes two

At least with depression we can be reasonably sure that negative emotional mood is one thing while weight loss is another. The confusion arises because it is not clear whether weight loss is a feature, a consequence, or merely a correlate of negative mood. However, the problem of definitional overlap can get even murkier. Sometimes we are looking at neither correlation *nor* causation. Sometimes what we have is conflation (Hughes, 2016). What might appear to be two separate variables – indeed, what might be extensively studied and reported in the research literature as two separate variables – are in fact not two variables but one: one variable with multiple identities.

Consider the case of social support and its relationship with stress. Social support attracts a lot of interest from researchers and policymakers alike. For nearly fifty years, it has been hailed as a psychosocial panacea capable of protecting us against physical and mental disease. But while depression is defined according to agreed diagnostic criteria, nobody can say with certainty what social support actually is. We have had various attempts to define it. Some psychologists subdivide it into subtypes (such as informational or emotional support); some distinguish between perceived and actual support; and some focus on separating its functional and structural aspects. But no one approach to defining social support has emerged as the consensus.

That said, a relatively small number of social support questionnaires dominate the research literature. One of the most popular is the Short Form Social Support Questionnaire (SSQ6; Sarason et al., 1987). In response to six questions, the respondent is asked to list the names of people they can rely on in various situations. Curiously, five of these six items make explicit reference to stressful life contexts, using phrases such as 'when you are under pressure', 'when you are feeling down-in-the-dumps', and 'when you are very upset'. These questions presuppose life stress. As such, only respondents who are accustomed to life stress should find these five questions pertinent. Respondents who never feel under pressure, are never down-in-the-dumps, and never get very upset ought to consider most of the SSQ6 irrelevant. Therefore, when asked to identify and list – by name – the people who help them when they are miserable, they should have no names to list, because they are *never* miserable. Even if they have large numbers of kind friends and family members, people who lead low-stress lives should report low social support on the SSQ6.

In fact, if you wanted a brief questionnaire to measure exposure to life stress, you might consider the SSQ6. People who are constantly under pressure, down-in-the-dumps, very upset, and so on, are likely to score high on this scale; people who experience little or no life stress are likely to score low. All told, you could find that the SSQ6 is a better measure of life stress than it is of social support.

Nonetheless hundreds of studies have employed the SSQ6 with the intention of exploring links between social support and stress. Many find that social support is rated as important by people who (in separate questionnaires) report life stress to be high. This should not be a surprise, given that the social support questionnaire is itself measuring life stress. The finding is that life stress is correlated with itself.

A similar situation arises with the study of positive emotion and life satisfaction. All that advice telling you to 'think positively' because it will lead to a better life owes its origin to studies linking these two variables. However, the questionnaires used to measure them bear strong resemblances to one another. The Differential Emotions Scale (DES; Izard et al., 1974), used to measure positive emotion, asks respondents to rate their feelings of 'Contentment'; meanwhile, the Satisfaction With Life Scale (SWLS; Diener et al., 1985), used to measure life satisfaction, asks respondents to rate their agreement with the statement, 'The conditions of my life are excellent'. Both scales are asking whether the respondent is content. If respondents report being content (or not) in one scale, then presumably they will also do so in the other.

The DES then asks respondents to rate their feelings of 'Gratitude'; the SWLS asks for a rating of 'So far I have gotten the important things that I want in life'. In this case, both instruments are asking whether the respondent is grateful. Again, highly grateful (or ungrateful) respondents should give corresponding answers in both scales. Where the DES asks respondents to rate their feelings of 'Pride', the SWLS asks them to rate 'If I could live my life over, I would change almost nothing'. Here, both instruments ask respondents if they are proud of themselves. Once more, people who are proud (or not proud) will say so both times. In fact, several items in the two scales resemble each other in this way. It is little wonder, therefore, that measures of positive emotion are so closely linked to measures of life satisfaction. Both instruments are measuring the same (single) thing.

In these cases, what look like two psychological constructs, one leading to the other, are actually just one construct, measured in two ways. In the first, we have a study where there are two measures of life stress – it's just that one of them is intended to measure social support. In the second case, we have two measures of happiness – it's just that one of them is labelled 'positive emotion' (happiness in general) while the other is labelled 'satisfaction with life' (happiness with life). These are not cases of confounding per se, because they do not involve two separate entities that intertwine, such as depressed mood and weight loss. Rather, they are cases of conflation, in which two allegedly separate themes are just different perspectives on the one merged entity.

In psychology the risk of conflation is high because of the inferential and abstract nature of psychological variables. As we cannot directly isolate and observe a thought or a feeling, we run the risk of interpreting questionnaire responses as evidence of underlying constructs that might not actually exist. It is easy to develop separate questionnaires, such as the DES and the SWLS, that produce distinct numerical scores even though they measure the same thing. The problem is that statistical data generated by such scales will be similar in form and structure to scores from questionnaires that measure *different* things.

There is nothing to warn (or stop) the researcher who insists on analysing such data as though there were two distinct variables to investigate. Psychologists are often advised not to confuse correlation with causation; conflation is arguably a much greater problem.

Towards replication: The problem of external validity

The third main type of validity is *external validity*. This refers to the extent to which an outcome seen in one setting (or group) resembles what would be found in other settings (or groups). In other words, does the outcome generalize? Sometimes a research study will lack external validity because it uses a sample of people who are not representative of the population at large. Other studies will lack external validity because they are conducted in artificial situations that are too far removed from the real world. In these cases, the research findings cannot reasonably be generalized. We will examine these problems in Chapters 5 and 6, respectively. Here, we will first mention the problem of low external validity in measurement. External validity in measurement can refer to the way a chosen method produces scores similar to those produced by other methods. It can also refer to the likelihood that a score produced in one setting will be produced again in another setting.

Judging a measurement method on its resemblance to an alternative might seem straightforward, but in fact it is more than a little complicated. For one thing, if the alternative method is effective (i.e., if it has good construct and internal validity), then why bother introducing a new method? Most likely, the extant approach is considered insufficient in some way. Researchers have concerns about its usefulness and wish to improve things by developing a new technique. They want to avoid the trap of falling back on old measures that have become obsolete or ones that were never that refined to begin with. Developing a new measure is a lot of work, so we should acknowledge the commitment of psychologists who take on the challenge. However, if the older measure is deficient, then it makes little sense to rely upon it as a quality control benchmark. And yet this is commonly done. A new scale or questionnaire is seen as having more merit if its measurements reflect the patterns of the old ones.

Secondly, as the target method is (presumably) not identical to the alternative that it is compared to, then it is unlikely there will be a perfect correlation between their scores. For example, were you to rank-order one hundred participants on their scores on a particular depression scale (for example, the BDI) and then rank the same participants using a second depression scale (say, the MDI), you will probably observe slightly different rankings for each scale. If you use one scale to judge the validity of the other, you will see that they are similar but not the same. But how then do you know which of the two is the more accurate? In other words, how do you use comparison to draw conclusions about validity?

External validity can also be judged on the basis of whether measurements are consistent across time and place. If an aptitude test has good external validity, then an able person should get high scores whenever and wherever the test is administered. This goes to the heart of psychology's replication problem. If a measurement method produces different scores on different occasions, then a study that includes that method will be difficult (if not impossible) to replicate.

Methods that vary are inherently deficient. If each time you step on your weighing scale it tells you that you are a different weight, you will probably feel it is time to purchase a new weighing scale. However, even if the scale gives you the *same* weight every time, this doesn't mean it is accurate. It just means it is consistent. It could of course be consistently *inaccurate*. This brings us to the final component of psychology's measurement crisis: poor measurement reliability.

Reliability: How steady?

Because of the abstract nature of psychological concepts, all psychological measurement will be subject to some retest instability. In other words, when a participant completes the same questionnaire or rating scale on multiple occasions, it is highly likely that their score will vary, just a little, from one occasion to another. The degree of variability can be quantified as a margin of error. If an instrument is to be any use, then it will be hoped that its margin of error is low. However, in psychology only in very few cases is this margin of error ever known.

Many IQ tests will include a facility to quantify and report the margin of error of an IQ score, and indeed it is generally considered poor form to report an IQ score *without* its associated margin of error. However, in psychology IQ tests are the exception rather than the norm. The vast majority of psychometric assessment instruments produce simple total scores, with no margin of error. Even though we know that there *must be* one, we just won't know what it is.

In some cases, the authors of a psychometric instrument – or maybe some other researchers – will attempt to establish its 'test-retest reliability'. This statistic sounds as if it might be a margin of error, but it isn't. Instead, it represents the average correlation between two sets of measurements taken from the same people at different times. In order to produce this information, the researchers will assemble a single sample of participants and arrange some way of administering the scale to them twice. When the data are in, the researchers compute the statistical correlation between the first set of scores and the second. Because it is a correlation, test-retest reliability is presented as a number between 0 and 1. The closer it is to a perfect 1.0, the better the reliability; and according to lore, test-retest reliability above 0.7 can be seen as acceptable. However, there are several problems with all of this.

Firstly, a correlation of 0.7 might seem chunky, but the relationship it represents is not as impressive as it looks. Correlations are statistics that reflect the closeness with which two sets of scores resemble each other. It is useful to note that what we call 'shared variance' of two variables is computed by squaring the correlation between them. This means that, for a correlation of 0.7, the overlap between the first and second sets of scores will be 0.49 or, more clearly, just below 50 per cent. So when an instrument's test-retest reliability is 0.7, this means that its scores on one occasion will predict its scores on a second occasion only half of the time. Any weighing scale that had this level of reliability would quickly find itself on the scrapheap.

It is true that what we consider an acceptable level of test-retest reliability might differ from variable to variable. Maybe it is unfair to compare a psychometric instrument to a weighing scale; after all, weighing scales are expected to be close to perfect. But this expectation for physics highlights the problem for psychology. What we consider acceptable often reveals the limits of our ambition. It is obvious that measuring weight is more straightforward than measuring, say, depression or intelligence. But this does not make 49 per cent shared variance any less problematic. A test-retest reliability that denotes fifty-fifty consistency between a first and second attempt to measure the same thing reflects a pretty low standard. It does not bode well for the replicability of any study that depends on it.

A second problem is that test-retest reliability does not account for logical reasons why scores may differ from test to retest. If participants complete the same psychometric scale twice, each experience of doing so may differ in ways that impact on responses. For example, performing an IQ test a second time might be affected by practice: a person's second score might represent a fair improvement resulting from having seen the test before. Indeed, in the case of just about any psychological measurement, the participants' memory of having previously completed the task could influence their second effort.

The experience of undergoing the first measurement may lead them to reflect on the subject matter in ways that changes the variable in question. For example, completing a depression questionnaire might lead to rumination that causes greater depression or to catharsis that relieves it. Further, the computation of test-retest reliability might be contaminated by the passing of time. If the interval between test and retest is too short, the participants might be tired or bored during the second sitting. If the interval is too long, the participants might be sufficiently older in ways that affect their scores. In short, test-retest reliability might be presented as a precise statistic, but it does not provide precise information as to why test and retest are not the same.

The biggest problem is that test-retest reliability is simply not the same as a margin of error. While it conveys information about variability, it does not tell us by how much an individual person's score might vary across repeated measurements, and therefore how much of a grain of salt we should be take when interpreting a score taken in a single sitting. As a correlation, a test-retest reliability statistic will simply reveal the degree of association between *sets* of scores, across *groups* of people.

In summary, any psychological measurement involves a margin of error. The problem is that in the vast majority of cases we simply don't know what it is. We sometimes have information on test-retest reliability, but the established conventions suggest that psychologists are remarkably accepting of inconsistency in measurement across time. They are not just happy with a weighing scale that measures something other than weight, they are happy for it to have a wobbly needle too.

Getting the measure of psychology's measurement problem

Psychology is a science reliant on the measurement of abstractions. As such it is reliant on attaining high-quality construct validity, internal validity, external validity, and reliability. The problem for psychological measurement is that none of these are ensured. Construct validity is frequently threatened by the elusive task of defining the intangible, the devilish first step in all psychological research. As well as requiring a consensus, concepts are often culture-bound and time-sensitive. The inferential nature of measurement means that the measure might be far removed from the concept being investigated. The concept being investigated might not even exist.

Internal validity is threatened by a tolerance of dubious control groups, which leads to false inflation of differences between groups, situations, and times. Confounding abounds in psychology research, and it seems to go largely unchecked.

External validity is often based on a less than satisfactory approach of benchmarking the supposedly new-and-improved measures against the admittedly old-and-deficient ones. And reliability, the basic requirement that a measurement produces a stable measure each time it is used, tends not to be tracked that closely. Where it is examined, the threshold for accepting low reliability seems strikingly forgiving.

Measurement in psychology is often presumed to be strong. Non-psychologists assume that psychologists know clearly what IQ is, how to measure internet addiction and stress, and where to draw the boundaries around things like positive emotion and life satisfaction. They also assume that, as scientists, psychologists would never employ a measurement method unless they knew how accurate it was. But these assumptions are misplaced. Some measurement methods are strong, but in some cases we know that measurement methods are weak. In other cases we don't know for sure whether they are strong or weak.

Many psychologists have turned psychometrics into a fine art, and there is a wealth of professional knowledge that can be deployed to the task. However, as with other sophistications, it seems that the bulk of psychologists have limited skills or awareness with regard to psychometrics. Some of this reflects a phobia about the statistical aspects of measurement. It is to psychology's statistical crisis that we now turn.

CHAPTER 4

'That Which Can Be Measured': Psychology's Statistical Crisis

The will of the people

'An election is a moral horror,' said a character in one of George Bernard Shaw's plays, 'as bad as a battle except for the blood' (Shaw, 1921). Be that as it may, bloodlessness seems preferable to the alternative. When citizens vote to shape the laws they live under, the avoidance of bloodshed elevates the enlightenment of the exercise. Effective elections save lives, bad ones cause social collapse. Political votes are central civic functions worthy of close attention and rigour.

The premise of elections is essentially that of scientific psychology. We employ standardized measurements to quantify people's wishes, lest they otherwise be misunderstood. The people's wishes, you see, are often unclear, influenced as they are by emotions, reasoning, opinions, loyalties, vested interests, cultural norms, and concentration levels. The sentiments of voters are exactly the type of abstract mental variable that psychologists deal with every day.

As such, the parallel effort to *predict* the electorate's desires through the use of political opinion polling is of particular psychological interest. In essence, political opinion polls constitute de facto psychology research studies. Pollsters are essentially behavioural scientists, gathering evidence using structured survey methods in an effort to calibrate the population's imminent choices. The factors that determine the quality of opinion polls are precisely the same as those affecting the merit of psychology research. Are the measurements effective? Do they have construct validity, internal validity, and external validity? Are they reliable? Are they interpreted correctly? And most importantly, do the findings translate into real-world actualities that are borne out by subsequent events? In other words, *are the polls replicable*?

We noted in Chapter 2 how citizens sometimes fret about the impact of fake news on electoral decision-making. Election news cycle are driven by coverage of opinion polls. If poll results are spurious and unreliable, then reporting them as otherwise could be seen as its own form of fake news. This is worrying because poll results are pivotal in shaping the electorate's intentions. They can create a vortex of runaway quasi-popularity for candidates who are deemed to be leading, turning an apparently foregone conclusion into a self-fulfilling prophesy. Or they can frighten an otherwise 'shy' subgroup of voters into supporting a candidate who appears to be falling behind. So likely is it that poll results will themselves proliferate as opinions, some countries outlaw them during campaign season. Other countries have toyed with banning them all year round (Chung, 2012).

A particularly difficult problem is the recurring feeling that political polls can be, well, highly inaccurate. An early example arose from the 1936 United States presidential election, for which the *Literary Digest* conducted what was then the largest opinion poll in history. Having sent the survey to over ten million people – one quarter of the electorate – the magazine declared that Alfred Landon would win the election by a landslide. As it turned out, it was his opponent Franklin D. Roosevelt who won the election by a landslide. It is hard to imagine a less accurate prediction in a two-candidate election than getting a landslide the wrong way round. Students of psephology often attribute the error to sampling: the *Literary Digest* identified its participants from telephone directories, thereby failing to reach the majority working classes whose votes ultimately determined the outcome. The debacle has become a comforting classic case study. It usefully reminds us how rarefied groups, such as the media, exist inside a bubble of privilege, which bursts upon exposure to the real world. The phone-owning classes will seem like an unproblematic sampling group for an election survey, if you believe that these are the only people who really matter (or vote). But nowadays, the story goes, pollsters know not to make such mistakes. Don't they?

Actually, right across the developed world, political opinion polls appear to be getting less and less accurate (Skibba, 2016). In recent years, several high-profile elections and ballots – in the United States, the United Kingdom, Canada, and France, to name but a few – have produced outcomes wildly at variance with what polls had predicted.

For the 2015 British general election, virtually *all* ninety-two opinion polls conducted during the six-week campaign were wrong (Cowling, 2016). Almost invariably, they had predicted a hung parliament, a result where no single party has a majority. By contrast, the actual election provided a clear majority to the leading government party. In 2017, there was a second election, giving the pollsters a chance to redeem their reputations. This time the vast majority of polls predicted a clear majority for that same party. The actual election then produced a hung parliament. It was as though the opinion polls were producing systematically erroneous results. Whatever the polls said would happen turned out to be the opposite of what *actually* happened.

The opinion polling industry is always keen to emphasize its focus on methodological rigour. Results are depicted in fine detail, replete with infographics and footnotes detailing statistical particulars. Large samples are boasted about, based on the view that maximizing a survey's reach is an all-conquering enhancement. As though imitating the precision of an electron microscope, 'margins of error' are reported alongside summaries of what respondents have said. The numbers seek to convey a sense of scientific confidence in the reported findings. They show how a range of opinion are covered and signpost the wisdom of claiming relative, rather than absolute, accuracy. In other words, the stats aim to convince readers that the poll results are *reasonable* – even if they are wrong.

It tells us something about the power of numbers as scientific devices that opinion poll sample sizes and margins of error are reported by journalists and devoured by readers and that in most cases neither understand their meaning or relevance in any way. In other words, few in the audience have a clue as to what constitutes a convincing sample size or how to compute or interpret a margin of error. This problem arises from the nature of behavioural science. It emerges throughout the whole of psychology, whether its research examines attitudes, behaviours, emotions, cognitions, abilities, or brain function.

In reality, these various figures are not as comforting as is frequently presumed. This is something of a recurring theme in psychology. Statistics are held in high regard even when they are misreported or abused. The replicability of psychology research is itself a statistical proposition: it relates to the quantitative similarity between one set of findings and another. If the statistical foundations of research are unstable, then replicability will eventually come crashing down. In psychology, the most important behavioural science of them all, the problems of iffy stats and unacknowledged researcher innumeracy run particularly deep.

The science of mind-reading

So what are pollsters (and other researchers) getting at when they cite sample sizes and margins of error? Sample size is mentioned for a few reasons. Firstly, the size of a sample can influence its representativeness. It is important that a sample includes space for all the different subgroups of the population that are relevant to the survey. However, this is much easier to discuss than to achieve. For example, we might not know exactly what relevant subgroups exist. While it makes sense to argue that larger samples have more potential to be representative, it is difficult to ensure that they fulfil this potential or to specify exactly what sample size is needed to do so. We will return to the challenge of representativeness in Chapter 5.

Secondly, sample size is also assumed to influence the statistical accuracy of results. It does this by affecting the second reported figure: margin of error. In the computation of margin of error, sample size is a denominator. Therefore, the larger the sample size, the smaller the margin of error. Common sense dictates that smaller margins of error are better; thus, larger samples must be preferable to smaller ones. But all this is more arithmetic projection than logical reality. Sometimes margins of error are bigger than you think. The 2 per cent margin of error that the Chinese government aims for when counting its population is bigger than the entire population of Australia (Golder, 2017). Other times margins of error are smaller than you think. Having a large sample might well reduce what mathematicians call 'error', but the practical difference made might be very limited indeed. In fact, insofar as their impact is liable to be misunderstood, larger samples might *promote* error. This is because, like many other methodological misunderstandings, they give researchers an unwarranted sense of security and an artificial belief in dodgy findings.

Margins of error are derived from statistics called 'confidence intervals', or CIs. For any statistic – that is to say, for any piece of information expressed as a number – it is possible to compute a CI. In short, the CI is a range of numbers deemed likely to contain the true value of the statistic in question. For example, a political opinion poll might find that Party A enjoys the support of between 40 per cent and 60 per cent of the population. In this case the CI is the 20-percentage-point space between the two numbers mentioned.

By custom, the margin of error is defined as occupying half the range of the CI. Therefore, if the CI is a 20-point gap between 40 per cent and 60 per cent, the margin of error will be the 10-point gap between 45 per cent and 55 per cent. CIs and margins of error will usually be described using the phrase 'plus or minus' (sometimes represented by the symbol '±'). Therefore, in media reports, our finding would typically be summarized as follows:

Party A's support is 50 per cent plus or minus 5 per cent.

CIs are computed by combining three pieces of information: the mean average of all the responses collected from the sample, the standard deviation of these responses, and the sample size. The mean average and the sample size are the easiest to understand. The mean average is the ordinary kind of average: the total of all scores divided by the number of scores. The sample size is what it says it is: the size of the sample, or the number of people who took part in the study or survey. The standard deviation is a little more complicated, but not outrageously so (e.g., it is taught to teenagers in high school). It is a number computed to represent the extent to which the actual scores are spread above or below the mean average. (Essentially, a standard deviation is computed by checking how far each score is from the mean average, squaring each of these distances, computing the mean average of these squared distances, and finally computing the square root of that latter mean average. A scientific calculator will do this with a few button presses.)

Once you have these three numbers, the formula for what we call a '95 per cent CI' is easy to apply:

$$CI = mean\,average \pm 1.96 \times \frac{standard\,deviation}{\sqrt{sample\,size}}$$

In plainer English, we might synopsize the formula like this:

> The 'confidence interval' is related to 'how spread out the scores are' divided by 'how big the sample is'.

In discussing CIs and margins of error, we need to bear in mind a number of important points. Firstly, a CI is not a guaranteed outcome. It is an estimate. We should always remember that samples are selected at random. This means that the views expressed to pollsters reflect a random subset of all the views in the population. This subset of views could be very idiosyncratic. It is therefore unsound to assume that the computed CI will accurately inform us about the population as a whole. For this reason, researchers hedge their bets by describing CIs not as certainties but as *relative likelihoods*. Most commonly, estimates are described as having 95 per cent likelihoods. This detail is not trivial and should really be incorporated into all summaries of results. However, it is rare to see this done in media reports. Were reporters actually to refer to this information, they might explain our finding like this:

> There is *a 95 per cent chance* that Party A's support is 50 per cent plus or minus 5 per cent.

But this phrasing is still not quite correct. Technically, the 95 per cent likelihood refers to the CI and not to the outcome (in other words, it refers to the 'plus or minus 5 per cent' and not to the '50 per cent'). It is intended to give us a sense of how often such margins of error will include the true value should the polling procedure be repeated multiple times.

This is based on a similar logic to the way we talk about coin-tosses. Most people will be aware that when a coin is tossed there is a 50 per cent chance of getting heads and a 50 per cent chance of getting tails. However, this does not provide any guarantee regarding a *single* coin toss. If you toss a coin once, you could get either heads *or* tails. If you toss the coin twice, you could get two heads, two tails, heads followed by tails, or tails followed by heads. The 50 per cent likelihood has little or no meaning if you toss the coin just once or twice. However, it has a lot of meaning if you toss the coin a large number of times. For example, if you toss the coin ten thousand times, then you can be reasonably sure that, overall, you will get *around* five thousand heads and *around* five thousand tails. You still won't be able to predict the order in which these heads and tails will come, but you can predict the overall totals of each.

In a similar way, the 95 per cent likelihood for a margin of error has little meaning if you conduct a survey or study just once or twice. Instead, its meaning only becomes tangible when the survey is repeated a very large number of times. If you do this, then *around* 95 per cent of your margins of error will contain the true value. Technically, if you could repeat the survey an *infinite* number of times, then *exactly* 95 per cent would do so.

Putting all this together, the description of our survey result would be much improved were it to be phrased like this:

> While we say support for Party A is 50 per cent plus or minus 5 per cent, the true level may fall outside this margin of error. We cannot be sure, because this is just one survey. What we *can* say is that if an *infinite* number of *identical* surveys were conducted, then 95 per cent of the reported margins of error *would* include the true level of support for Party A (although we wouldn't be able to tell which ones).

As well as being quite a mouthful, such a statement is philosophically challenging. In the real world, people's voting attitudes will be influenced by survey results they hear about in media reports. Were it even possible to conduct an infinite number of surveys, each successive poll would be biased by the published results of the previous one. Therefore, in order to meet the requirement for an infinite number of *identical* surveys, each survey should be conducted in secret. Only then could we correctly interpret the margins of error. But that still leaves the problem of infinity. By definition, we would never have enough time to conduct an infinite number of surveys.

The point is that when margins of error are presented in published opinion polls, they reflect very subtle points of theoretical mathematical reasoning. Even in the best possible circumstances – where survey takers succeed in avoiding all methodological imperfections – the results will be nowhere near as persuasively accurate as might be implied by terms such as 'margin', 'confidence', or '95 per cent'.

Even more problematically, margins of error cover only one error at a time. They are rendered uninterpretable should a second error emerge.

The word 'error' in the term 'margin of error' refers to the error associated with sampling: the extent to which views of any randomly selected subset of the population deviate from those of the population as a whole. CI estimates assume that sampling error is the *only* source of error in the data. Any other shortcoming in the survey – such as poorly phrased questions or a non-representative sample – effectively negates the validity of any stated CI or margin of error.

Therefore, the footnote to our survey result would be further enhanced were it extended to include the following:

> Please note that our previous statement about an infinite number of identical surveys is based on an assumption that there are no shortcomings in the conduct of the surveys. If, for example, respondents are *not* representative of the population, then we can make no statistical statement about Party A's true support.

Of course, a more succinct reduction of everything would be:

> If – and only if – the people who took part in the survey are *right*, then Party A has around 50 per cent support. But we don't really know.

A survey finding phrased like that would sound far from impressive. But it would be much more accurate than the disclaimers appearing in most media reports about opinion polls.

From description to explanation

When considering behavioural research, it is useful to remember a general rule: statistical results are never as specific as they might appear. This is because all results are subject to margins of error (a point habitually glossed over in the psychology research literature). However, as we have now seen, it is useful also to remember a second general rule: *margins of error* are never as specific as *they* might appear. They are probabilistic statements that may or may not apply in a given case, and they are dependent on assumptions that may or may not ever be true in reality.

We can note that these complexities apply when we are measuring a single psychological construct, such as political preference. Using numbers to summarize a single concept involves description rather than explanation. For this reason we call such summaries 'descriptive statistics'. However, in behavioural research we are often interested in explaining the connections among *multiple* different concepts, such as whether X affects Y, or whether X differs from one group of the population to another. To do this, we use arithmetic to work out the statistical patterns of all the variables and to assess how probable it is that these patterns systematically reflect each other. We call this use of numbers 'inferential statistics', because we are making inferences. We are drawing conclusions from our statistical evidence.

By and large there are two broad categories of inference (although in research they are often intertwined). The first category is where variables or concepts might *associate*. This occurs when two concepts appear to be interlinked or for some reason to follow similar or synchronized patterns. We might conclude that one of the concepts is the cause of the other. An example is where we find that tobacco smoking rates appear to be interlinked with heart disease rates. When people smoke a lot, they get more heart attacks; when they rarely smoke, they remain relatively disease-free. In this case we see that smoking and disease are associated; we might even conclude that one causes (or partially causes) the other. In psychology, many statistical approaches are used to examine these types of links. The most commonly mentioned will be 'correlations' and 'regressions'.

The second category of inference is where variables or concepts might *differentiate*. This happens where concepts emerge in different ways in different circumstances, such as when a single concept changes over time or when it manifests itself differently in different groups of people. An example is where we find that smoking rates appear to be higher in men than in women. In this case we use statistical approaches to draw a conclusion about whether there is a meaningful gender difference in smoking. Because we are considering differences instead of associations, it is not straightforward to employ correlations or regressions. Instead we use a category of statistical techniques that examine the differences between groups of numbers. The most frequently mentioned of these will be 'analysis of variance'.

The nuances concerning margins of error apply in every case, because every research study involves the measurement of variables (and thus the specificity of such measurement). However, inferential statistics bring their own additional nuances. These largely concern the issue of probability, which is effectively the process of estimating how likely something is to be true. The philosopher David Hume once wrote that 'all knowledge degenerates into probability' (Hume, 1798). Perhaps it is unfortunate then that so many human beings find the description, handling, and calculation of probability to be truly mind-boggling. Psychologists, whose professional life's work depends on such things, are no exception.

Studying individuals by looking at groups

When pollsters attempt to gauge the electorate's voting intentions, they sometimes focus on the preferences of different groups in the population. They may speculate about divergences between rich and poor, urban and rural, or young and old. These distinctions all make sense. Rich and poor people will be affected differently by how politicians manage the tax and benefits systems. Urban and rural communities will have different stakes in decisions about infrastructural development. Young and old people are likely to have different levels of involvement in services such as healthcare or education. It makes sense to be interested in the different voting preferences across these various groups. Much psychology research takes such a form: examining samples of different groups of people and then assessing whether or not some psychological variable – for example, attitudes, preferences, personalities, behaviours, depression rates, or reaction times – differs from one group to another.

In any psychological research study, the analysis of group differences will follow conventional lines. Firstly, the researchers will attempt to conceptualize the variable of interest as numerical scores, one for each group. To illustrate, imagine the variable of interest is people's approval of Party A, rated on a scale of 0 ('I don't like Party A') to 10 ('I love Party A'). The researchers will seek to quantify average approval of Party A among different groups in their sample (e.g., urban people and rural people). Almost certainly, approval ratings will be numerically different, even if only slightly. The next step will be to conduct a statistical test to see if this observed numerical difference is meaningful in real-world terms. Maybe it is so small as to fall within the margin of error. Alternatively, maybe it is large enough to be genuinely informative.

There are a number of caveats to bear in mind when proceeding with all of this. An initial concern is that confidence in the findings will be undermined if the measures lack validity or reliability (see Chapter 3). There will also be problems if the study sample is not representative of the population as a whole (such as in the *Literary Digest*'s 1936 US election survey). If we satisfy ourselves that there are no serious measurement or sampling issues, then the next set of caveats comes into view. These all relate to statistics.

First, it is often the case that numerical scores are poor at depicting what the researcher wants to study. The problems here relate to the way we simplify complexity. Take mean averages. You may be familiar with the example of a community in which the mean average number of children per family is 1.5. Obviously, no actual family will have 1.5 children. However, this is *so* obvious that few readers would feel misled by receiving information in this form. They would consider it reasonable. But not everything is as evident. In the same community, the average family income might be £55,000. Maybe one family, or even several, have exactly this income. But then again, maybe *no* family has: maybe the average is inflated or deflated by a small number of families who have unusual incomes. In this case, the use of a mean average might be considered *un*reasonable: readers might feel *very* misled were they to receive information in this form. Put simply, the mean average of any set of figures helps give us a feel for those figures, but it does not inform us about the details.

Statistically-minded psychologists often warn that averages should not be considered in isolation. Because of problems like fractional children and outlier earnings, it is never sufficient to consider means on their own. Most often it is recommended that we take standard deviations into account. You will recall that standard deviations depict the general 'spread' of data. The smaller the standard deviation, the more the data lie close to the mean average. The larger the standard deviation, the more the data are spread out. This can help us detect whether a mean average family income is a good depiction of what most families earn.

However, while standard deviations add information, they do not complete the picture. Contrary to popular belief (popular among psychologists at least), it is possible for two sets of data to be quite different from each other and yet still have the same mean average *and* standard deviation. Take, for example, the following sets of scores from a fictional experiment:

Group A: 7, 8, 9, 10, 11, 12, 13, 14, 15

Group B: 10, 10, 10, 10, 10, 10, 17.25

In Group A, nine participants have each scored differently on a test, with their scores ranging from seven to fifteen. In Group B, all but one of the seven participants have the same score (10), while the outlier has scored 17.25. In both groups, the mean average score is 11, and the standard deviation is 2.74. From the point of view of how psychology data are typically summarized in research (i.e., as means and standard deviations), these two groups are identical. However, when you look at the individual data, you can see that they are absolutely different.

Ultimately, the use of means and standard deviations is a very incomplete way of encapsulating data. At the very least, it raises the prospect of ambiguity. Groups that are different can be confused for groups that are the same. This is not a trivial problem for psychology research (or for opinion polls), given how much researcher time is spent evaluating group differences. A related concern is more conceptual. If psychology is a science in which we study individuals, then our most common way of handling data in research is poorly matched to our ambitions. Rather than examine individuals per se, our research conventions involve the clumping together of groups, followed by the scrutiny of generalities about these groups. The individual variations within groups are concealed, rather than revealed, by our methods.

Blind tests

A second caveat is related to this problem of clumping and its role in statistical analysis. Here we will consider the way psychologists analyse differences, but we can note that the same logic applies when they analyse associations. As mentioned above, when psychology researchers examine group differences, they will most likely use a procedure known as analysis of variance, or ANOVA. Sometimes they will use a test called a t-test. The t-test is effectively a quick way of doing an ANOVA when the research question is simple. (It was developed by a Dublin brewery chemist, who wanted to evaluate differences in beer quality resulting from different production methods.) Both procedures assess whether two or more sets of data are similar to each other.

Most psychology undergraduates will be familiar with the chore of entering all the numbers to be analysed into a spreadsheet or software package and then either applying a mathematical formula by hand or getting a computer to run the procedure automatically. Ultimately, the analysis will lead to a p-value that can be used to determine the status of the group difference. As described in Chapter 1, the custom is that p-values less than 0.05 are declared 'statistically significant'. In the present case, this means that when p is below 0.05, the group difference is interpreted to be an actual difference rather than a fluke.

When students first learn how to conduct an ANOVA or t-test, the process can appear more complicated than it really is. We can tell why when we look at one of the formulae:

$$t = \frac{\overline{X}_1 - \overline{X}_2}{\sqrt{\frac{n_1 - 1s_{x_1}^2 + n_2 - 1s_{x_2}^2}{n_1 + n_2 - 2}} \times \sqrt{\frac{1}{n_1} + \frac{1}{n_2}}}$$

You can note that this is the formula for the t-test. In other words, it is intended to be the *shortcut*. The full native ANOVA uses the following:

$$F = \frac{\sum_{i=1}^{k} n_i \overline{Y}_i - \overline{Y}^2 / K - 1}{\sum_{i=1}^{K} \sum_{j=1}^{n_i} Y_{ij} - \overline{Y}_i^2 / N - K}$$

These multi-storey formulae are intuitively bamboozling to most people. The main reason they are so cumbersome is because they compute everything required from the so-called raw data (the numbers relating to each single participant). The impression formed when analysing data this way – or when entering raw data into a computer – is that the test derives p from all that detail. However, this is not really the case. The procedure to determine p requires just means, standard deviations, and sample sizes. Most of the detail in the formulae – and most of the work done by software – is required simply to compute these means and standard deviations from the raw scores.

To compare two groups, all an ANOVA (or *t*-test) really requires is *six* simple numbers: the mean average score for each group, the standard deviation of the scores for each group, and the number of participants in each group. This is true no matter how large the study is. If there are five thousand participants in each group, there will be ten thousand scores to enter into the spreadsheet. That is certainly a lot of manual data entry. But once the data are entered, the software (or formula) will crunch all ten thousand numbers down into six pieces of information. It will then use those six bits to tell us what *p* is.

As described above, it is possible for two quite different groups of scores to produce the same mean and standard deviation. In such a case, the typical statistical test will be unable to differentiate them. As far as an ANOVA or *t*-test is concerned, the groups will be identical. They have the same vital statistics and so, regardless of underlying variability, will be considered indistinguishable. Tests are blind to vital details: they read headlines, not prose.

The significance of significance

The biggest problem with statistical testing, however, is knowing what to do about significance. Statistical significance is widely treated as a benchmark of credibility in research. As outlined in Chapter 1, psychology researchers go to great lengths to analyse and then re-analyse their data in order to isolate findings that breach the 0.05 threshold, transporting them to the promised land of significance. It is as though statistical significance in one's findings determines one's personal significance as a human being.

As described in Chapter 1, psychologists habitually exploit Rampant Methodological Flexibility to squeeze their studies so hard that, eventually, *p*s less than 0.05 pop out. Unfettered confirmation bias means that such squeezing often amounts to little more than culpable misdirection. Researchers don't discover significance, they *make* their findings significant. Even minor adjustments to sample sizes, scoring protocols, or basic research procedures will have knock-on effects on statistical results.

In nearly every situation a way can be found to pare that p-value down to the desired level. This activity has become known as 'p-hacking'. In a famous example, Simmons, Nelson, and Simonsohn (2011) reported a study in which otherwise elusive significance was produced by excluding a third of the sample from analysis and then incorporating a plethora of peripheral measures (as 'control variables') in the statistical tests. Their purpose was to demonstrate the impact of methodological flexibility on statistical significance. By constructing their analyses so deliberately, they were able to attribute significance to a particularly implausible finding. In short, they showed that participants who were *randomly allocated* to an experimental treatment, in which they listened to a Beatles song, became, as a result, *statistically significantly younger* than participants allocated to a different group. With poker faces but tongues firmly lodged in cheeks, the authors said this must *prove* that listening to certain types of music reduces your actual chronological age.

We have already seen that statistical tests, such as ANOVA, are blind to the details of datasets. We can add that they are oblivious to the contents of research. Unreasonable findings can quite plausibly reach statistical significance, because statistical significance is a matter of arithmetic, not reason. However, this does not mean that statistical tests are useless. The reasonableness of research reflects the researcher who conducts it, not the statistical test they use. If a statistical test can be cited as 'proof' that music reverses human ageing, then this is not a problem with statistical tests. It is a problem with the way researchers interpret (or conduct) their studies. That statistical tests are blind should not cause difficulty. In a rational world, the human researcher is the one who does the seeing.

Which hypothesis?

It is extremely important that scientists be competent at conducting – and interpreting – statistical tests. However, for many years there has been widespread concern that psychology researchers are poorly informed about statistics and particularly inept at interpreting statistical significance. Judging by their behaviour and utterances, it would appear that most simply do not know what 'statistical significance' really means. This defect in understanding is non-trivial. It leads to the widespread dissemination of groundless conclusions and contributes directly to psychology's replication problem.

In psychology, the conventional statistical practice is known as 'null hypothesis significance testing', or NHST. For any single study NHST involves the drafting of two possible hypotheses, or research predictions. The first is called the 'main' hypothesis (sometimes the 'experimental' hypothesis). This is essentially the answer to the study question the researcher has in mind. For example, if the research question is 'Are women more extraverted than men?', then the main hypothesis will be 'Yes, women are more extraverted than men'.

The second hypothesis is called the 'null' hypothesis (the hypothesis that gives NHST its name). This is a nullified version of the main hypothesis. Note that the null hypothesis is not the *inverse* or *negative* of the main hypothesis, merely a nullified form of it. Therefore, in the aforementioned example, the null hypothesis would be 'No, there is no gender difference in extraversion' (rather than, for example, 'No, men are more extraverted than women'). NHST tests the null hypothesis. If the associated p-value is less than 0.05, then the null hypothesis is said to be 'rejected'. In other words, it is considered (albeit tentatively) untrue.

In reality, there is no absolute need to compose a null hypothesis in words. It is possible to perform the arithmetic of statistical testing without having done so. Because of this, it is no surprise that psychology researchers can be adept at finding shortcuts through their work. Given there is no absolute need to prepare a null hypothesis in words, many researchers will simply think up their *main* hypothesis and then proceed straight to statistical analysis. Null hypotheses are treated as theoretical notions rather than practical ones. Insofar as they are thought about at all, they are associated with long-forgotten undergraduate psychology statistics courses, not real-world research contexts.

Many psychologists dismiss the null hypothesis as a quaint pedagogical affectation. Some bemoan it as truly irksome and roll their eyes at the poor methodological pedant who happens to ask about it. To most psychology researchers it seems simply obvious that the focus should be on the main hypothesis. After all, this is the putative answer to the central research question, the implicit prediction of the study, the *purpose* of the entire exercise.

The problem is that understanding p-values requires, at the very least, a broad understanding of what statistical tests are trying to do. And what statistical tests are trying to do is tell us about null hypotheses. Therefore, a proper understanding of p is unlikely to emerge if null hypotheses are seen as quirks or oddities.

The results of statistical tests never 'support' particular hypotheses or findings, no matter how much researchers state the contrary. This is because null hypotheses and main hypotheses are seldom monogamous. While your null hypothesis is indeed the null of your main hypothesis of which you are aware, it is also the null of a large number of other main hypotheses to which you remain oblivious. Take our example of the null hypothesis 'No, there is no gender difference in extraversion.' If a statistical test tells us that this null hypothesis should be rejected, then this implies that it is unlikely to be true. But if it is untrue, then what is true? Well, it could be that our main hypothesis – 'Women are more extraverted than men' – is true. But that is not very specific. Maybe a different version ('Women are vastly more extraverted than men') is true, while another ('Women are slightly more extraverted than men') is false. Equally, it could be that a diametrically opposing hypothesis ('Men are more extraverted than women') is true. Or maybe the reality is multi-conditional ('Before the age of 40, women are more extraverted than men; after that there is no difference'). Or perhaps a more tentative main hypothesis ('There is a gender difference in extraversion, although we cannot be sure of the direction') could be said to be true, without there being sufficient information to go into any more detail.

The point is that all these different versions of the main hypothesis correspond to a single null hypothesis – namely, 'No, there is no gender difference in extraversion'. Based on the examples in the above paragraph alone, at least five different researchers could start five different studies with five different predictions, and yet all could base their statistical test on the same null hypothesis. If this null hypothesis is rejected, then which main hypothesis is supported? Which researcher wins the prize for having made the most astute prediction? Note that if they are the type of psychologists who go straight from main hypothesis to statistical test *without even considering what the null hypothesis might be*, they won't even realize that their conclusions are interchangeable.

What p means (and what it doesn't mean)

Based on several surveys, we can confidently say that most psychologists – a clear majority – just don't know what p-values mean (Haller and Krauss, 2002). By implication, when discussing statistical significance with psychologists, you are more likely to hear an incorrect explanation of p than a correct one. Ask a group of them what p-values represent, and most of the answers you hear will simply be wrong.

For instance, all of the following interpretations of p are commonly presented in the literature. All are considered correct by large numbers of psychologists whenever they are surveyed about them. However, all of them, without exception, are wrong:

- p is the probability that the main hypothesis is false.
- p is the probability that the null hypothesis is true.
- p is the probability that rejecting the null hypothesis is the wrong decision.
- p is the probability of observing a different result in a future replication of the study.
- p is the probability that the observed results are a fluke.
- p is the probability that the observed results are due to error (e.g., sampling error or measurement error).

Not only do the majority of psychologists in general mistakenly believe one (or more!) of these statements, but so apparently do the majority of academic psychologists who teach statistics and research methods to students (Haller and Krauss, 2002).

The p-value produced by NHST is not any of those things. Rather, it is best expressed as something like this:

- p is the probability of gathering the obtained data in a world in which the null hypothesis is true.

Note that p is the probability of data, not the probability of hypotheses. It refers to the likelihood of gathering the particular data you have gathered should the null hypothesis be correct. For example, consider again the example of a research study into the matter of gender differences in extraversion, the null hypothesis of which is 'No, there is no gender difference in extraversion'. Imagine you have gathered data that show that women have a mean extraversion score of 5.63 and men have a mean extraversion score of 4.72. Your statistical test will then return a particular p-value. This p-value represents *the probability of women scoring above 5.63 and men scoring below 4.72 in a world where there is no gender difference in extraversion.* That is all.

If p is high, we could logically conclude that, in a world with no gender difference in extraversion, group mean scores of 5.63 and 4.72 are unremarkable. It implies that although the scores are numerically different, they are not *substantively* different. The numerical difference is just part of the normal fluctuation of naturally occurring numbers.

On the other hand, if p is low, we can conclude that the scores are highly *im*probable. The arithmetic difference between 5.63 and 4.72 is meaningful and unlikely to be part of the normal noise of natural numbers. If this is our conclusion – literally, if we conclude that it is highly improbable to observe such scores in a world where there is no gender difference – then we might begin to question the merit of the null hypothesis. In other words, we might question whether it is reasonable to imagine such a world where there is no gender difference in extraversion. When p is low, the evidence makes us *feel* as if the null hypothesis is unlikely.

That is the logic of NHST: we use statistical tests to examine the probability of particular data, and from this we form opinions about the relevant null hypothesis. But this is entirely a matter of human judgement. Rationality is required. The p-value produced by our statistical test does not provide any quantitative information about our hypotheses. It certainly does not specify how probable our main hypothesis is. It fails to remove the risk that replication might reveal the purported finding to be spurious. And, as intuitively tempting as it might appear, the logic of using the probability of data to form opinions about hypotheses, even the null hypothesis, is simply unsound. We shall now see why.

The limits of NHST

Sometimes psychologists believe that describing p as the probability of observed data is simply a roundabout way of saying it represents the probability of the null hypothesis. In other words, they believe that the probability of the data is equal to that of the hypothesis. They think that the two are interchangeable. However, they really aren't. In short, *the probability of the data (given the hypothesis)* is entirely different from *the probability of the hypothesis (given the data)*.

Explanations involving hypotheses and data are always going to be difficult to think through, so let us consider an alternative example (see Branch, 2014). Imagine you are about to have a meeting with a visitor from China. You do not yet know who the visitor is. In your daydream you speculate about one day meeting the Chinese president, although you know well that it is extremely unlikely that he will visit you. Still, daydreams are interesting, so you set about trying to figure out the likelihood that this will happen. What you are trying to compute is the following:

> The probability that I will meet the president of China (given that I am about to meet a Chinese person).

As there are over 1.3 billion Chinese people, the likelihood that your Chinese visitor will in fact be the president is extremely remote. For one thing, the president of China doesn't just go around visiting people unannounced. But even if your visitor was a randomly selected Chinese person, the probability would still be less than 0.000000001. It is a probability more remote than your brain can truly comprehend.

Now let us imagine that you really *are* going to meet the president of China! Perhaps you are an important head of state, and your country is holding some kind of political summit. This time your daydream involves thinking about the Chinese system of government and the requirements to be president. One requirement is that the president of China must be Chinese. Therefore, given that you are about to meet the Chinese president, you can be assured that you will meet a Chinese person. This daydream involves the following computation:

> The probability that I will meet a Chinese person (given that I am about to meet the president of China).

In this case, the probability is extremely high. In fact, given the rule that the president of China must be Chinese, it is an absolute certainty: numerically, the probability is 1.0.

This could hardly be more different from the probability arising from your previous computation. In other words, *the probability that you will meet the president of China (given that you are about to meet a Chinese person)* is utterly different from *the probability that you will meet a Chinese person (given that you are about to meet the president of China)*. In exactly the same way,

The probability of the data (given the null hypothesis)

is absolutely different from

The probability of the null hypothesis (given the data).

And this is why the *p*-value produced by NHST provides no information whatsoever about the reliability of hypotheses in a study. The *p*-value is the probability of the data, given the null hypothesis; not the probability of the null hypothesis, given the data. The two are simply not interchangeable.

When you obtain statistical significance, you can by all means attempt to comment on the viability of the null hypothesis, but you do so at your own peril. The *p*-value might show the probability of your data to be 0.000000001, but this does not mean that the probability of your null hypothesis is 0.000000001. The probability of your null hypothesis could be 1.0. We simply cannot use the *p*-value of one to specify the probability of the other. The humble *p* simply does not tell us what most psychologists claim it does.

There is one final issue to incorporate into the explanation of *p*. It is reminiscent of what we said above about margins of error and CIs. Namely, NHST is a reliable approach only insofar as its technical requirements are met. The mathematics work only if various arithmetic assumptions hold true.

All statistical tests have assumptions. They all assume that samples are drawn randomly. Most also make assumptions about the distribution (or spread) of data within study datasets, assuming them to be similar across the groups that are compared. Many assumptions are far from intuitive; some are incomprehensible to the uninitiated (to illustrate, try to imagine what a lay person would make of such requirements as 'homoscedasticity', 'sphericity', or 'homogeneity of regression slopes').

Computations of p are unreliable when assumptions are not met. And there is no guarantee they ever will be.

In short, if p is low (e.g., if $p < 0.05$) then this could mean that the data are improbable given the null hypothesis. But it could also be a sign *that the assumptions of the statistical test were breached*. To account for these scenarios, our description of p should extend to resemble the following:

- p is either (a) the probability of gathering the obtained data in a world in which the null hypothesis is true, or (b) a quasi-random by-product of having violated the assumptions of statistical tests (e.g., assumptions concerning measurement, sampling, homoscedasticity, etc.). It is impossible to tell which of (a) or (b) holds true in any given case.

It is bad enough that psychologists tend to be unaware of what p really means. It is worse that most are not just ignorant but labour under treacherously false beliefs. Statistical significance does not provide information about the reliability of hypotheses. This means that it does not tell us whether effects are replicable.

Nonetheless, statistical significance effectively remains a prerequisite for publication of research in the psychological sciences (Branch, 2014). In essence, decisions to publish research depend on a widespread but mistaken belief that the truth of hypotheses can be demonstrated statistically. The myth that p describes the truth of hypotheses exacerbates the replication crisis, because it discourages researchers from conducting replications (Branch, 2014). With no logical foundation, statistical significance has become a proxy for replication. But NHST, while reasonable and logical in its own terms, can never achieve what replication is intended to do.

Misunderstandings outweigh understandings

Here is another factoid to dampen enthusiasm around p. With NHST (which is, remember, the standard statistical approach in psychology), all results become statistically significant if a large enough dataset is used. That's correct. All of them. So if your data show a small gender difference in extraversion that is not significant right now, just collect more data and re-conduct your analysis. Should the same thing happen again, go get even more. Repeat the process for as many iterations as it takes. In theory, if you keep enlarging your sample, the *p*-value of the gender difference will get smaller and smaller and smaller (the direction of the gender difference might flip back and over, but don't worry about that). Eventually *p* will fall below 0.05, and you will have a statistically significant finding. It is a mathematical certainty.

There are a number of reasons for this. Firstly, statistical tests are extremely pedantic. Remember: they are blind and oblivious. As such, they consider anything 'not identical' to be 'different'. Secondly, the ability of a statistical test to do its job is proportionate to sample size. With every additional person in the sample, the more powerful the analysis becomes and the better it is at noticing very small differences. But thirdly, and most importantly, statistical tests assume noise-free data. They believe in order and exactitude in nature. Even if the difference between two groups is minuscule, a statistical test will see it as having the same potential for significance as any other difference. Ultimately, the only way to avoid significance would be if the numbers involved were exact replicas, complete to as many decimal places as can be recorded. If both men and women had mean extraversion scores of 5.0, a statistical test would declare the groups identical. However, if men had a mean of 5.0000000001 and women a mean of 5.0000000002, then even this tiny difference would be heralded as statistically significant if the sample was sufficiently large.

One implication of this is that, in an organic universe, the null hypothesis is always false. In nature, noise simply does not cancel itself out. The prospect that large samples of men and women would have identical mean scores for extraversion – to an infinite number of decimal places – defies what is credible. It would be akin to discovering two duplicate meadows in the countryside in which every single blade of grass in one is matched by a corresponding blade in the other, occupying the same position and possessing the same shape, size, and colour as its counterpart. In the natural world, carbon copies are mythological. Even identical twins are not identical. The fact that NHST assumes the possibility of true non-difference means that it is looking for something that cannot be found.

It is only margins of error that allow NHST to declare entities to be 'the same' as each other, when blurred vision prevents true detail from being seen. As sample sizes increase, margins of error decrease, and with that the acuity of NHST improves. With a sufficiently large sample, the sensitivity of NHST is enhanced to the point where it can detect even the smallest detail of difference and, lacking common sense, declare it 'significant'.

It may be that millions of participants are required to reach this point of senseless statistical significance. However, the principle holds true in probabilistic terms no matter what the size of the study. In other words, the *larger* a sample, the *more likely* it is that small differences will be flagged as significant findings. In some cases, these small differences will be meaningful (such as a small difference in cancer cell growth). In other cases, they will be trivial (such as a difference in extraversion scores of 0.0000000001). In yet other cases, they will just be random; organic background noises seeping into naturally occurring data. Yet again, we see that statistical significance is just a very poor guide to the reliability or value of findings.

Size matters

To be fair, methodology textbooks in psychology have long warned students not to overemphasize statistical significance. 'Statistical significance is not the same as practical significance' is a common mantra. What matters is not how probable the data are (given the null hypothesis). What matters is the size of the effect that has been established. The misuse of p is regularly discussed in blog posts, workshops, and conferences. Various scientific bodies and journal publishers urge authors to focus on 'effect sizes' more than significance. Indeed, for decades, psychology journals have featured more-than-occasional articles demeaning NHST as a logically perilous misadventure. Many of these have become citation classics. Jacob Cohen, the renowned psychology statistician, published his landmark paper on the topic over two decades ago (Cohen, 1994). Its title famously ridiculed p-values:

'The Earth is Round ($p < .05$)'.

Even then, Cohen was at pains to point out that his dismissals of NHST were unoriginal. According to him, they were at least forty years old when he wrote his paper back in the early 1990s.

In 2015 the prominent journal *Basic and Applied Social Psychology* banned the use of *p*-values from its pages (Trafimow and Marks, 2015). In 2016, the American Statistical Association issued a statement intended to address 'the prevalent misuses of, and misconceptions relating to, *p*-values' and to usher in a 'post $p < 0.05$ era' (Wasserstein and Lazar, 2016). It seems as though psychologists have been knocking, mocking, and defrocking NHST for the best part of half a century.

Effect-size statistics aim to supplement *p*-values by providing information not just about probability but also scale. If there is a gender difference in extraversion, it is insufficient alone to know that it is statistically significant. An effect-size statistic will tell you how *big* the difference is. Of course, size is relative. As alluded to above, a tiny gender difference in mean extraversion might not have any meaningful implication, whereas an equivalent difference in cancer cell growth could be very important. We also need to take account of the fact that differences *between* groups are often inconsequential when compared against differences *within* groups. Nonetheless, size clearly matters, and information about the size of effects seen in research is critically important. And yet, as outlined in Chapter 1, the identification of statistical significance – not effect size – remains the basis on which most research is selected for publication in psychology. It is a key element of the dynamic that produces psychology's replication crisis.

Statistical limitations as human failings

It is something of a conundrum that, in several ways, advances in technology have made psychology's statistical problems worse rather than better. Firstly, it is now much easier to conduct large-scale research and to accumulate those very large samples that make trivially small effects statistically significant. Large-dataset research is more common now than in the past. On balance, it is true that large samples are preferable to small ones. The typically tiny samples used in psychology research – especially brain imaging research – remain a major cause for concern.

However, when p-values are prioritized more than effect sizes, the downside of large-scale research becomes conspicuous. The temptation to consider trivial effects significant without considering their real-world impact becomes irresistible.

A good example of this problem is the field of gender differences research. Several cases relate to the availability of large-scale datasets in educational testing. By now hundreds of studies have examined potential gender differences in test performance across a number of academic subject areas. Many have focused on mathematics. While plenty of statistically significant gender differences have been reported, most are utterly puny in terms of real-world relevance.

For example, in a sample of final year high school students in the United States young men were found to have statistically significantly better mathematics scores than young women (Hyde et al., 2008). However, the effect size for this difference was very small. Using a convention known as Cohen's d, the effect was determined to equate to a d of 0.06, which is extremely tiny indeed. It implies that less than 0.1 per cent of the gender difference was due to gender per se. In other words, there was a near complete overlap in men's and women's maths scores.

The main reason the effect was statistically significant was sample size: the dataset contained scores from nearly half a million students. While the gender difference was statistically significant, it was of little or no actual significance in real-world terms. In fact, in all likelihood the difference was not just trivial but random. When the authors performed equivalent analyses for different age-cohorts of students, they found similarly sized statistically significant gender differences throughout the system, but with *no consistent direction*. For example, while nine-year-old boys outscored nine-year-old girls, ten-year-old girls outscored ten-year-old boys. The directions of the various (minuscule) gender differences were roughly equal across the cohorts, with no apparent association with age.

It is remarkable that, even in the face of robust empirical evidence of their incidental triviality, gender differences in maths performance have remained the target of ever-durable stereotyping. Underrepresentation of women in the upper echelons of academic mathematics and engineering is regularly explained away by reference to supposed gender differences in numeracy. The irony that such stereotypes are themselves founded on deficient mathematical reasoning is of little consolation.

A second unfortunate consequence of advances in technology relates to the automation of statistical analysis. Before software was readily available, psychologists were required to compute analyses by hand or, if they were lucky, with the aid of a pocket calculator. Even then, the process of producing p-values was indirect. A statistical test would only allow a researcher to compute an indicator variable (e.g., a t-test would compute a variable called 't'). The researcher would then use this indicator variable to look up p in a statistical table, most likely one included as an appendix in a textbook. Often, the resulting p-value was not an exact number but merely an approximation or a range of possible values (for example, '$0.01 < p < 0.02$'). Manual statistical analysis was so labour-intensive that researchers were highly selective in the analyses they chose to conduct.

In contrast, modern statistical software is user-friendly and quick, and greatly reduces the cost in time and effort required to perform even complicated analyses. Therefore there is little to discourage researchers from conducting many different tests to examine many different hypotheses or from pursuing endless subtle variations of each one. This exacerbates the problem of Rampant Methodological Flexibility. It makes it easier for researchers to run ever more idiosyncratic analyses until such time as a desirable p is produced. In addition, given the probabilistic nature of statistical testing, the conducting of multiple analyses necessarily increases the likelihood that random (i.e., false-positive) findings will emerge. Automation allows a researcher to analyse an exhaustive list of combinations and permutations of variables and, through a dredging process, to identify randomly significant effects that can be reported as though the researcher knew in advance what to look for.

And finally, because statistical tests are blind and oblivious, there is nothing inherent in the software to prevent an incompetent researcher from conducting analyses that are incorrectly framed or that use datasets that violate the assumptions of the tests being deployed. In other words, the user-friendly automation of statistical analysis allows researcher ineptitude to flourish undetected in a way that was far less likely when actual (human) numeracy was required.

The truth about statistics

The cliché refers to 'lies, damned lies, and statistics'. A famous book purports to tell us *How to Lie with Statistics* (Huff, 1954). This theme warns us to be wary of numbers and to fear their inherent dishonesty. It reinforces a scepticism towards statistical information and, by extension, towards science itself. While frequently presented, the stereotype is extremely unhelpful.

This is because the solution to the problem of bad numbers is better numbers, not no numbers. The idea that statistics are a tool for lying to people hides the fact that it is people who do the lying. In reality, readers should not be sceptical of statistics any more than they should be sceptical of verifiable facts and logical reasoning. Statistics are essentially numerical versions of these things. Quantitative data are numerical versions of 'verifiable facts', while statistical analysis is a numerical version of 'logical reasoning'. When information is presented in the form of numbers, its truth-likelihood increases.

The problem is that many readers, and many reporters (be they researchers or journalists), simply don't appreciate what the underlying truths are. Audiences are lied to by messengers who wish to lie to them. Sometimes they are misled by messengers who misunderstand the message. But they are not presented with a dishonesty inherent in the message itself. Thus, audiences will be protected by more statistical literacy rather than less and by better statistical information rather than none at all.

It is often observed that innumeracy is one of the few socially acceptable forms of ignorance. People regularly admit to being terrible at maths with a comfort that would never apply were they to confess to being unable to spell. If anything, innumeracy is depicted as an endearing trait or a symptom of cuteness, whereas illiteracy is seen as intolerable. The apparent acceptance of arithmetic ineptitude is seen not only in social attitudes and relationships but also in politics, business, and bureaucracy. As described by one columnist in the *Financial Times*, 'Employers take a dim view of anyone who makes spelling mistakes on a CV, yet they routinely hire people who can hardly count' (Kellaway, 2017).

In a very similar way, while psychology produces guidelines that promote competence in statistics, in practical terms psychologists regularly downplay the impact of innumeracy among researchers. This can be seen in the balkanization of skills within academic psychology departments. Nearly all such departments contain within their membership a small subgroup who are known to be statistical gurus. Competence at statistical analysis is seen as a finite specialism, in the same way that cognitive psychology or developmental psychology are seen as specialisms. Rather than *all* researcher psychologists being expected to be statistically numerate, it seems professionally acceptable in most cases simply not to be. When it comes to analysing actual data in actual research, the average psychologist will make the best fist of it they can, notwithstanding their limited understanding of what to do. Alternatively, as is increasingly the case, they will outsource the job to a statistically-minded collaborator, on a no-questions-asked basis.

This soft spot for innumeracy has negative effects on the field. When journal editors and reviewers run scared from statistics, liberally giving the benefit of the doubt to whatever researchers have written, the prevalence of iffy stats in the literature expands perniciously. The favouring of narrative over numbers might make for more newsworthy research results, but it does little to ensure that the results are actually replicable. The widespread misunderstanding of p-values, the prevalence of p-hacking and data dredging, and the prioritizing of statistical significance (rather than effect size) as a criterion for publication are all very damaging. The use of NHST as a substitute for replication is utterly ill-conceived and virtually guarantees the emergence of a replication crisis in psychology.

CHAPTER **5** # 'We Are The World': Psychology's Sampling Crisis

What women want (according to a sample of women)

Love is all around. However, that doesn't mean it is the same every-where. As we all know, love comes in many different forms. It is dangerous to generalize. But one sweeping statement we can offer is that love – or at least sexual love – has existed for a very long time. Sexual love has been a constant throughout human history. After all, with-out the instinct to procreate (or, at least, the urge to attempt to do so), the human race would not have gotten very far in evolutionary terms. Even when intellectualized as poetry or institutionalized as marriage, love is at root a visceral impulse. We love because we carry genes that predispose us to seek partners to mate with. We love because we are the descendants of organisms that succeeded in the competition for biological reproduction. We love because we are, ultimately, animals.

The idea that our prehistory can influence our present-day sexuality is a popular fascination. It is a theme extensively covered by news media. Every day there are reports to describe yet more ways in which psychologists have shown our sexual behaviour to reflect our troglodyte ancestry. Often these involve signs or signals that we emit to subconsciously tell potential mates that we want to pair up with them. Maybe our body smell will show ideal partners that we have compatible genes. Maybe the way we purse our lips will signal our trustworthiness to prospective suitors. Maybe the colour of our clothes will highlight when we are optimally fertile.

However, there are problems with this type of research. One is the fact that evolutionary processes, by their very nature, preserve features that will have been useful to our ancestors, but not necessarily to us. Because the vast majority of our forebears lived before humans realized that it might be nice to actually cook meat before eating it, you can take it there is much about our physiologies that is nowadays obsolete. Our ability to choose a suitable mate by smell (even if it were significant) would not necessarily do us any good. It might have made our children a bit more resistant to prehistoric bacteria, but we are unlikely to see such benefits today. Those millennia are long gone.

A second problem is more mundane. It relates to the making of so much from very little, the drawing of grand conclusions from incomplete evidentiary particles. Evolutionary hypotheses involve a huge amount of extrapolation, across thousands of years, with very little by way of direct evidence to go on. The filling-in-of-blanks involved in such research often requires the pooling of information from fragmented or distantly connected sources. Underlying evolutionary influences are inferred from present-day behaviours and testimonies. These are unlikely to be truly comprehensive.

Take for example the (frequent) claim that women are more likely to wear certain colour clothes when fertile. Such hypotheses are based on the fact that many animals signal fertility by changing colour. Egg-laying ants display different body patterns compared to infertile ants (Liebig et al., 2000). Swordtail fish have 'brood patches' on their abdomens that change size when they are ready for courtship (Benson, 2007). Striped plateau lizards develop similar patches on their throats that intensify in colour during ovulation, thereby stimulating males to initiate inter-course (Weiss, 2006). And in Gibraltar Barbary macaques, females have flaps of perineal skin that change size in proportion to fertility, allowing attentive males to 'focus their reproductive effort accordingly' (Brauch et al., 2007).

Human females tend not to have such tell-tale skin blotches, brood patches, or flesh bulges. In fact, empirical evidence for fertility signal-ling in humans has eluded researchers for decades. In 2013, a group of Canadian psychologists decided to settle the matter once and for all, with a survey (Beall and Tracy, 2013). They asked some 'regularly ovulating' women to report on their menstrual cycles. They then asked them what coloured shirts they were wearing. Thinking of everything, they asked women whose shirts were multi-coloured to classify their garment in terms of its 'most prevalent' hue. And that was pretty much it. The data suggested that women wearing red (or pink) shirts were more likely be between six and fourteen days post-menstruation. As the research-ers considered this timeframe to be one of 'high conception risk', they concluded that their signalling hypothesis was supported. As far as they were concerned, women wear red when fertile. (Not that the researchers wished to jump to conclusions: they noted that 'the underlying mechanism accounting for the present findings is, as of yet, unknown'; p. 1839.)

This study attracted high-profile media interest. Several newspapers and websites reported the finding. The result has become something of a legendary factoid, and it is now regularly cited in *Did you know?* articles concerning the science of sexuality. The overall idea has implicit appeal. It resonates with our culture's belief that red is the colour of romance, as well as research showing that men find red attractive. But the study was not without criticism. Some commentators questioned its definition of fertility. Others accused the researchers of exploiting Rampant Methodological Flexibility to produce a significant finding in the absence of a compelling a priori hypothesis. But the biggest shortcoming was more fundamental, a problem that afflicts thousands of psychology studies. The problem was the extremely limited sample of humanity upon which the researchers saw fit to make extravagantly grand claims.

In attempting to isolate a human equivalent to the fertility signalling observable throughout the animal kingdom, the authors investigated the self-reports of two groups of North American women. The first group comprised one hundred women who completed an online survey. The researchers did not meet these women face-to-face. Most serious scientists acknowledge that it can be difficult to verify whether participants in online surveys are who they say they are. Having real-life respondents lessens concerns about the authenticity of survey data. Therefore, the researchers sought to supplement their study by recruiting some women college students from their own university. Of these they attracted a grand total of twenty-four.

So there it is. The study based its analysis on 124 women, all from North America, just twenty-four of whom participated in person. The journal article reporting the research provided no information about these women other than their ages. There were no details about their race, occupations, socioeconomic status, urban/rural backgrounds, marital status, or sexual preferences. Other psychologists wishing to replicate the study would know what questions to ask, but they would not know whom to ask them of. Those seeking to apply these findings to the wider human population would struggle to know exactly which humans they could generalize to.

Human evolution is a species-wide phenomenon, not a geopolitical one. Therefore, when the authors invoke evolution as the basis for their hypothesis, we must assume they believe their finding to be of universal relevance. In other words, all women, everywhere, wear red when fertile (or, at least, are statistically more likely to). It is nonsensical to suggest that only Canadians and Americans would have inherited this fertility-signalling impulse through evolutionary processes. So, all told, the researchers infer a species-wide effect – a heretofore unacknowledged behaviour now said to be exhibited by literally billions of people – on the basis of asking 124 of them, in one region, about their clothes.

In short, the authors' assumptions were simple: when it comes to signalling willingness for sex, North American women who attend college and/ or participate in online surveys are (a) generally all the same, and (b) identical to all other women who have ever lived, throughout the entire evolutionary lineage of humankind. You hardly need to be an expert on sampling to recognize the limitations of this approach.

The inconvenient truth about convenience samples

In theory, a research study should scrutinize a sufficient range of cases to bolster a confident declaration about the subject under investigation. A survey of political voting intentions should include an assortment of likely voters, so that diversity in opinion can be included and assessed. Any such survey that focuses on a narrow sample of like-minded individuals will, by definition, fail. A study of human reaction times should not rely on examining just one person, because one individual's reactions might be unusually slow or fast. It would make much more sense to measure reaction times in a group of people, ideally one large enough to provide a sense of what 'people in general' are like. And – surely – an investigation into biologically primed sexual impulses in pre-menopausal women should include a wide range of such women, broad enough to encapsulate the attitudes, belief systems, sartorial cultures, sexual histories, socioeconomic statuses, marriage traditions, and colour-blindness levels seen across the whole of humanity.

The theory of sampling is to draw together a selection of cases that reflect what you want to know about the population in general. Therefore, a sample should feature sufficient representativeness to facilitate generalization. Selecting participants at random (or something close to random) is ideal, because people who rush to volunteer for studies do not necessarily share the attributes of those who don't. In essence, random selection is best because it affords all members of the population an equal chance of being included. Over time and on average, this will meet the aspiration to faithfully represent the entire population in the samples of research in general. Such, at least, is sampling in theory.

Of course, as in life, practice typically falls short of theory. This is because of several pragmatic limitations. One is the impossibility of ensuring that all members of a population really do have an equal chance of being selected. For example, if your study concerns the linguistic abilities of three-year-olds, you simply cannot provide equal participation opportunities to every three-year-old on the planet. Going local seldom helps, because a database of every three-year-old in a given country, or even in a given city, is unlikely to be available. Most likely, your attempt to recruit three-year-olds will lead you to approach specific institutions, schools, or clubs to whom parents have entrusted their infant offspring. But the more you restrict geographical scope, the less confident you can be that your findings will apply to people who exist outside that scope. If you find that three-year-olds in a particular school in Birmingham are able to compose six-word sentences, can you be sure this ability will be present among three-year-olds from other schools (never mind from other cities or countries)?

A second pragmatic barrier is that not all members of the population will be _able_ to participate in research. Some might be excluded because of socioeconomic disadvantage. Others may lack a sufficient level of education, such as an ability to read, whereas some may not even hear about the study because of where they live. Still others might be incapacitated by illness or disability. Such problems greatly hamper the representativeness of samples, especially when research targets these very issues. For example, research on the impact of depression will be very incomplete if those people most debilitated by the disease are unable to participate. A related problem affects longitudinal studies where some participants drop out over time or even die. Studying only those who last the course can leave researchers with false impressions about how people are affected by change or intervention.

A third problem is that not everybody _wants_ to participate in psychology research. Believe it or not, even among the able-bodied, the economically enriched, and the well-educated, there are people whose passions are simply unstirred by the prospect of being a researcher's guinea pig. People who are shy, who are sceptical, or who are antisocial are most unlikely to lend their services to research. Such apathy will surely undermine investigations into, respectively, behaviour, attitudes, and personality. Research on sexual behaviour often finds that people are refreshingly, if not surprisingly, tolerant of progressive norms. However, this may be because such studies do not hear from sexually conservative folk, who decline to volunteer for surveys that they know will concern such subject matter.

Sometimes researchers argue that randomly sampling participants from the entire global population is simply impractical and forcing people who do not want to participate to do so anyway is simply unethical. Undoubtedly both observations are correct. Researchers have little choice but to approach participants who are approachable and to recruit those who are recruitable. But just because common practice falls short of theoretical standards does not make the theoretical standards less important. Nor does it magically infuse the common practice with robustness.

When Freud concluded that women suffer from penis envy, we should remember that he did so on the basis of a small number of nineteenth-century case studies involving middle-class women in Austro-Hungarian Vienna. That such a theory seems not to hold true outside that particular context should come as no surprise. Similarly, when Maslow developed his 'hierarchy of needs', in which personal self-actualization (rather than communitarian harmony) is said to be the ultimate target of human motivation, he did so by interviewing college students in California in the 1950s. It is little wonder that his theory seems not to apply to today's Americans and probably never did to citizens of other, less individualistic, societies. In short, poor sampling leads directly to poor replicability. Practical constraints offer no excuse for bad practice.

The sliver effect

It has occasionally been observed that most prominent psychology journals could usefully be renamed. For example, the *Journal of Personality and Social Psychology* could become the *Journal of Personality and Social Psychology of American Undergraduate Psychology Students* (Henrich, Heine, and Norenzayan, 2010). This would reflect the fact that more than half its articles feature research on such students (Arnett, 2008). As it happens, half of the articles that *don't* feature students simply feature other Americans instead. US psychologists have long been concerned that the major American journals focus too much on the study of their fellow citizens (Arnett, 2008). The worry is that too much is made of research focusing on a country where just 5 per cent of the world's population live. Given that most of this research examines a socioeconomically homogeneous subset of US society, the problem is probably much more acute than initial appearances suggest.

But non-American research does little better in capturing the diversity of human beings. Even when psychology examines other nationalities, it tends to examine groups who *resemble* those socioeconomically homogeneous American samples. Across the main international journals of psychology, over 95 per cent of research articles involve participants from North America, Europe, Australia, or Israel, regions that represent just 12 per cent of people in the world (Arnett, 2008). For this reason, psychology's favourite participants have been described as 'WEIRD': their backgrounds are Western, Educated, Industrialized, Rich, and Democratic (Henrich, Heine, and Norenzayan, 2010). Psychology draws grand conclusions about human beings by examining just a sliver of humanity.

Research does not always transfer well from place to place. This is especially true for research into cultural concepts such as language. It is perhaps obvious that studies of language acquisition, comprehension, and teaching might not easily generalize from one linguistic community to another. For example, studies of how English-speaking children learn to read or write will not be particularly informative when considering how Chinese-speaking children learn to do so.

English-language script has an alphabet; in Chinese language is perceived in terms of logograms (characters that represent entire words, such as '人' ['rén'] which means 'person'). It is worth bearing in mind that Chinese is a native language to over one billion people, whereas English is native to fewer than 400 million. We might therefore consider Chinese to be the more valid example of how human beings psychologically code language into writing. If most research on children's reading ability is conducted within a linguistically peculiar subset, then psychologists will have a poor basis for making inferences about how children in general learn to read.

Of course, the effects of language differences are not confined to scripts and sounds. They also involve semantics. Variations in language sometimes augment variations in how people socially convey their thoughts. For example, English speakers are immersed in a language that directly communicates a contrast between singular and plural, but Chinese speakers don't make this distinction in quite the same way. To a Chinese speaker, the statement '我吃肉' ('Wǒ chī ròu') can indicate either of the following:

- I eat meat (right now).
- I eat meats (of different kinds).

In addition, given the logographic origin of the verb '吃' ('chī'), the statement can refer to several paraphrastic meanings other than 'to eat':

- I am receptive to eating meat (I am not a vegetarian).
- I subsist on meat (and nothing else).

Such semantic differences complicate the task of using Western research to draw generalized conclusions about how humans verbally interact. The way in which English speakers convey their thoughts – and, critically, how they interpret or anticipate what other people say to them – does not provide a basis on which to describe all humans.

In fact, many linguists consider English to be somewhat quirky. Notwithstanding its widespread usage, it is quite an *unusual* example of a human language. For example, in English one of the commonest ways to ask a yes/no question is to use what is called 'word-order inversion'. In other words, where a declaration might be 'She is Chinese' or 'She is beautiful', the corresponding yes/no questions will be 'Is she Chinese?' or 'Is she beautiful?' Switching the order of 'She is' to create the question 'Is she?' comes naturally to English speakers from a very early age. It reflects such a common device that most native speakers consider it a point of *logic* rather than a purely linguistic convention. However, despite this appearance of common sense, such use of word inversion is extremely rare across human languages. In fact, fewer than 2 per cent of the seven thousand human languages known to exist use this method to convert declarations into questions (Majid and Levinson, 2010).

One way this directly affects psychology is when psychologists use English-language norms as a basis on which to posit internal cognitive realities in human minds. For example, for almost a century child psychologists have conducted literally thousands of studies into the way infants comprehend concepts such as _in_ and _on_. They claim to have demonstrated that little babies can understand such spatial relationships even before they have learned what the words mean. The vast bulk of this research is conducted from an English-language perspective: the implication is always that English-language prepositions reflect an inherently human way of thinking.

However, spatial relations are conceptualized and categorized very differently in different languages. In Spanish the same word is used to express 'in' and 'on' (therefore, 'El libro está _en_ la caja' can mean either 'The book is _in_ the box' or 'The book is _on_ the box'). In Dutch different words correspond to 'on' depending on the where the objects are located in space (thus, 'De foto staat _op_ de tafel' means 'The photo is _on_ the table', while 'De foto staat _aan_ de muur' means 'The photo is _on_ the wall'). In Korean the distinction between 'in' and 'on' is not used. Instead, there are prepositions for 'tightly locked with' or 'loosely locked with' (see Bowerman and Choi, 2001). Whereas English speakers place 'an apple _in_ a bowl' and 'a cup _on_ a table', Korean speakers consider these to be the same category of preposition (where objects are loosely touching each other). While English speakers put a 'CD _in_ its case' and 'a cap _on_ a biro', Korean speakers consider _these_ to be the same category of preposition (where objects are tightly interlocked with each other).

So when psychology research claims that prelingual children can understand the differences between _in_ and _on_, such a finding must be seen as somewhat idiosyncratic. The children's inherent abilities to distinguish spatial relations most likely exceed what the English language alone can describe. Presumably children can intuit Korean-style relations as well as English-style ones, were the experimenters to test them accordingly. The use of English-language prepositions as a template for human infant cognition is highly arbitrary (Majid and Levinson, 2010). It tells us more about the psychology of the researchers than it does about that of the participants being researched.

Differences in languages can even lead to differences in how people foster social cohesion. English speakers are familiar with modal verbs (e.g., 'could', 'would', 'should'). These are auxiliary words that give nuance to another verb. Chinese does not feature such constructions, which has implications for how nuance is conveyed. Chinese speakers who learn

English often find modal verbs odd. Therefore, instead of saying 'Could you open the door?' a Chinese speaker who learns English might be more comfortable simply saying 'Open the door!' To the Chinese speaker this statement reflects how Chinese is spoken and so makes perfect sense. To their English-speaking listener it sounds a tad peremptory, if not indeed aggressive. In English the use of modal verbs allows for a particular type of etiquette that is missing in Chinese.

Of course, this cuts both ways: English speakers say many things that correspondingly strike Chinese speakers as rude. Etiquette in everyday life is an important aspect of how people build relationships. If languages differ in how they deal with etiquette, then a linguistically narrow evidence base will hamper efforts to explain some of the most interesting nuances of human interaction.

In short, geographically restricted research can suffer from very poor generalizability at the level of language. The challenges relate not just to superficial disparities in vocabulary or writing but also in the nature and use of semantics for everyday exchange, as well as the context and communication of politeness, decorum, and cohesion. Psychology research conducted in Western countries does not easily shed light on how people interact with each other in other places.

Culture beyond language

Clearly, language differences will interfere with the generalizability, and thus replicability, of much psychology research. However, language is not the only psychological feature to vary from place to place. Several psychological concepts are so embedded in culture as to render psychology's reliance on WEIRD samples highly questionable. There is substantial evidence to show that social support, love relationships, social influence, conformity, altruism, cooperation, fairness, self-esteem, reinforcement, and many other topics studied extensively in psychology are intricately intertwined with cultural norms and variation.

It has long been known that even visual perception is subject to cultural effects. Psychologists originally assumed visual perception to be a basic process, one that is species-specific and hard-wired into all human brains. However, this has been disproven in a number of studies. A significant body of research has shown that shape perception is heavily influenced by early-life experience. Childhood exposure to contours, corners, and intersections provides a type of optical calibration that determines a person's perceptual skills in later life. In standardized visual perception tests, people from industrialized societies tend to perform differently when compared to people from other places, perhaps reflecting their immersion in environments where engineered products and infrastructure are ever-present (Henrich, 2008). Vision research conducted on WEIRD populations cannot be assumed to have universal relevance.

Culture also greatly influences basic motor skill acquisition and, thus, our ability to understand and navigate space. Western psychology textbooks often cite classic studies in which babies are coaxed to crawl over a 'visual cliff' (they are actually protected by an invisible glass floor). The aim of the research is to establish the age at which infants begin to detect depth and thus fear falling (Gibson and Walk, 1960). The original studies showed that while a handful of babies up to the age of about a year glided gaily across the glass, most balked at the cliff edge and refused to budge. The researchers suggested that this showed human babies to be able to comprehend depth by the time they can crawl. But although crawling in babies is very familiar to Westerners, it is actually a learned behaviour shaped by cultural immersion. In some societies crawling emerges much earlier than in the West because of the way parents interact with their infant children. In other places crawling is significantly delayed because infants are made lie on their backs for most of the day. In some cultures crawling is frowned upon altogether. Babies skip crawling, only moving around by themselves when they learn how to walk (Karasik, Adolph, and Bornstein, 2010). Given this diversity in motor-skill milestones, whether or how the visual cliff findings would generalize to non-WEIRD babies is simply unclear.

Even brain imaging research is hampered by a focus on WEIRD participants. We saw in Chapter 1 that neuroimaging studies tend to rely on tiny samples. This is partly because the relevant technology is so expensive. Researchers must limit its use. Another consequence is that imaging research is usually confined to countries that can afford to provide advanced infrastructure to their scientists. Well over 90 per cent of peer-reviewed brain imaging studies come from Western countries (Chiao, 2009).

It is often assumed that all brains are alike; however, relevant investigations have shown marked cultural differences in brain activity. For example, when processing objects (Gutchess et al., 2006) or thinking about oneself or one's family (Zhu et al., 2007), Western participants show brain activity patterns that are somewhat distinct from those of other people. This may well reflect how habits of cognition (such as whether to focus on objects in complex visual scenes) deviate as a function of cultural reinforcement, rather than there being inherent cultural differences in brain architecture. Nonetheless, the fact that cultural differences manifest themselves in measured brain activity makes it difficult to justify a near total reliance on WEIRD populations to provide participants for neuroimaging research.

In short, most psychology research is based on skewed, unrepresentative samples. With rarefied living conditions, homogeneous culture, and a singular language, the participants most studied by psychologists are in some ways among the *least* representative of human beings (Henrich, Heine, and Norenzayan, 2010). They comprise a minority who experience a life quite unlike that of the majority, with characteristics of environmental context that are discernible in unique social attitudes, behavioural outcomes, and brain functions.

The problem of homogeneous sampling is not a political shortcoming or a failure by psychologists to be democratic. Rather, it is a scientific flaw that affects the merit of the field. The purpose of psychology is to establish an evidence-based understanding of human beings, to extract general principles that can account for people's thoughts, feelings, and behaviours. A corollary is the endeavour to produce findings that are replicable, an effort that is greatly undermined by homogeneous sampling. Psychologists can hardly set out what it means to be human if they study but a sliver of humanity.

Mental health as cultural compatibility

Maybe WEIRD populations are not so 'weird' after all. That they comprise a minority of the human race is indisputable. However, there are at least two senses in which this minority might nonetheless represent a pure form of the human condition. Firstly, it is useful to remember that the species *Homo sapiens* is part of a taxonomic family of primates that includes four extant genera: orangutans, gorillas, chimpanzees, and humans. Of these, orangutans, gorillas, and (most) chimpanzees exhibit biological propensities for autonomy, mobility, low sociability, and individualism. In other words, our fellow great apes do not typically live in collectivist societies where, because of economic necessity or history, there is pressure to conform to social mores and standards. Maybe we humans are more like these great apes than we appreciate.

The second point is that for over 95 per cent of its history, the human species lived as hunter-gatherers. We must assume therefore that the bulk of human evolution was shaped by an existence in which people foraged independently for food. The combining of effort across large agricultural communities in order to cultivate plants and animals came very late within this evolutionary timeframe. Such observations have led a number of psychologists to suggest that WEIRD societies – which place high value on autonomy, mobility, individualism, and freedom of choice – are in fact a better 'fit' for the biologically evolved nature that human beings are likely to possess (Maryanski, 2010).

This does not alter the fact that psychology's focus on WEIRD populations restricts its ability to establish the full scope of human thoughts, feelings, and behaviours. But it does draw attention to the way human adaptation is shaped as much by cultural surroundings as it is by biological predispositions. What we often refer to as mental health might alternatively be seen as a question of compatibility. Mental health is effectively the extent to which people find themselves suited to their environments in a given place, at a given time.

The renowned psychiatric philosopher Bill Fulford was alluding to this when he described mental illness as a 'failure of "ordinary" doing' (Fulford, 1989; p. 120). He was describing the need to take context into account when considering whether a person's pattern of thoughts, feelings, or behaviour might constitute a mental disease. 'Ordinary doing' is the process of existing in day-to-day situations; the failure of ordinary doing occurs when there is a divergence between what it is a person does and what it is that is situationally appropriate. In other words, mental illness is not just a function of the person; it is also a function of the person's surroundings.

Distinctions between mental health and ill-health are intertwined in situational norms. Behaviours that are typical in one place can be atypical elsewhere. For instance, cultural norms for emotional expression vary widely around the world. In some countries it is considered desirable that people express their emotions vividly and loudly. Public crying is seen as highly appropriate, and people may even ululate in front of others when anguished. Lamentation of this kind draws sympathy from the community and is an important feature of social coping. However, other cultures have completely different norms. In many societies, it is considered highly *inappropriate* for people to express their emotions in elaborate or conspicuous ways. In these places, where forbearance and stoicism are highly valued, it is expected that people restrain their emotions during difficult circumstances, even when grieving. A person who goes so far as to howl in public would be considered extremely bizarre, and their neighbours would be more likely to be scared than sympathetic. This propensity to violate the social norm of stoicism could lead others to worry about the person's mental health. It could even bring them a diagnosis of a mental illness.

For decades it has been suspected that members of cultural minorities might frequently receive inadequate psychiatric diagnoses. For one thing, exclusionary attitudes can impede them from encountering services where their difficulties might be detected. There is also the ever-present concern that clinicians can be culturally insensitive, or even racist, and so less diligent in their diagnostic efforts (Sashidharan, 2001). But it is undoubtedly the case that poor diagnoses often result from physicians' failure to appreciate intercultural difference.

In mental health, it is common for minorities to be assessed using the norms of majorities. Immigrants are assumed to be indistinguishable from indigenous patients, and minorities in general are treated the same as the mainstream (Lopez and Guarnaccia, 2000). When cultural differences are under-appreciated, it becomes more likely that culturally specific behaviours will be deemed deviant. This bias might help explain the inflated levels of mental health interventions frequently seen in minority communities. For example, in Britain black and minority ethnic people are disproportionately more likely to be detained under the Mental Health Act (Singh et al., 2007). Such problems not only affect clinical practice, they also affect research. If the bulk of mental health research is conducted on majority populations (as is typically the case), then its applicability to minorities is immediately called into question. If findings cannot be generalized, it should be no surprise that research cannot be replicated.

Similar issues arise when attempting to extrapolate mental health research from one country to another. Mental health constructs are so embedded in cultural norm-setting that the expression of mental disorders is often geographically specific. This specificity is such that some recognized mental health conditions occur only in certain societies. Examples of culture-specific syndromes include Zar (a Middle-Eastern condition where people laugh and cry as if possessed by demons), Taijin kyofusho (a Japanese social phobia where people fear that others will be disgusted or offended by their presence), and Koro (a disorder seen in some South-East Asian countries, in which men believe that their genitals are shrinking).

It has also been suggested that several major disorders seen in mainstream Western psychiatry reflect experiences of industrialized populations and are not so common elsewhere. Conditions such as Post-Traumatic Stress Disorder and Attention Deficit Hyperactivity Disorder are frequently diagnosed in North America and Europe but are relatively rare in other places. The existence of culture-specific mental health conditions highlights the link between social context and psychiatric well-being. In mental health research, as in the rest of psychology, WEIRD populations are inherently unrepresentative.

Many of the commonest culture-specific syndromes have been recognized for several decades. A number were first documented by anthropologists in the nineteenth century. In recent years a different, but related, concern has arisen. This is the fact that even those mental health disorders that have been considered generic may themselves vary in expression from place to place.

One example is anorexia nervosa. Anorexic people (usually women) suffer significant weight loss and become dangerously thin. They engage in extreme behaviours to limit their calorie intake, forgoing most food and possibly using laxatives or forcing themselves to vomit whenever they do actually eat. These actions are assumed to result from distortions of thought. According to formal diagnostic guidelines, people with anorexia are unable to properly appreciate their body shape. They believe themselves to be fat and deny the seriousness of their emaciation. Furthermore, they experience an intense fear of gaining weight.

Anorexia has been recognized in Europe and North America for over a century. However, outside Europe and North America anorexia can look quite different. For example, in Hong Kong women exhibit the emaciation and food-avoidant behaviour seen in the West, but the psychological aspect of their condition is not the same. In Hong Kong anorexic persons typically recognize their emaciation. They also do not appear to fear being fat, nor do they wish to lose weight in order to become more attractive. While the evidence base is sparse, clinical reports suggest that Hong Kong women with anorexia engage in self-emaciation and food-avoidance for reasons other than body-image concern (Watters, 2010).

If such culture-specific differences are borne out by rigorous research, then this would have significant implications for our understanding of anorexia. Perhaps unsurprisingly, formal criteria for psychiatric diagnoses reflect how mental health conditions are experienced in Western countries. The criteria for anorexia are based on cases typically seen in these places. However, the version of the condition seen in Hong Kong casts significant doubt on the long-held belief that anorexic food-avoidance results from disordered patterns of thought. It seems possible that anorexic women in different societies develop varying after-the-fact rationalizations for their behaviour, most likely influenced by local cultural attitudes towards body shape and puberty.

Rather than *causing* anorexia, perhaps body-shape denial and pathological fear of fatness are culturally determined *consequences* of the condition. If so then, with respect to anorexia at least, psychology needs to radically rethink its entire approach to diagnosis, intervention, and mental health promotion.

Without sampling, we are nothing

The potential of psychology is undoubtedly enriched by the sheer diversity of human beings. However, as we discussed in Chapter 4, psychology research often involves the scrutiny of groups of individuals, rather than individuals per se. The one-size-fits-all approach implicit in such statistical methods is mirrored by the use of culturally narrow samples in psychological science. Simply put, psychology gives undue attention to a monocultural sliver of humanity, and from this draws unwarranted and extravagant claims about the propensities, proclivities, and perceptions of people in general.

It is true that psychologists frequently draw attention to the impact of culture on psychological development. For example, the nature–nurture debate, epigenetics, and cross-cultural psychology are readily recognized as mainstream concerns. But culture is often a much broader notion than is typically appreciated. It encapsulates not just language and geography but also how people see their world, how they navigate their surroundings, and how they experience and address the vicissitudes of modern life.

Despite drawing attention to the empirical importance of sampling diversity, the fact remains that psychology recruits most of its research participants from backgrounds that are Western, educated, industrialized, rich, and democratic. These societies account for barely one-tenth of the human race. In attempting to build an understanding of people in general, psychology scrutinizes samples from which 'people in general' are largely excluded.

The cultural narrowing of psychology highlights both prospects and problems for the field. On the positive side, it shows how future widening of the research sampling frame will help not only produce a more comprehensive evidence-based account of the human experience but also directly test hypotheses relating to the role of cultural influences on well-being. On the negative side, it draws attention to the parlous state of things as they are right now. Sampling theory requires that studies scrutinize representative selections of instances of whatever it is that is being studied. Nonrandom skewing of how samples are selected undermines the logic of generalization and impedes the replicability of research.

The preponderance of participants from culturally homogeneous and geographically concentrated populations makes it impossible for psychology to address human diversity to anything approaching an adequate extent. This not only hampers research, it also endangers any attempt at evidence-based practice, be it in the realm of education, healthcare, social policy, or psychiatry. In understanding how some people adapt and thrive, we mislead ourselves into believing that we know how to assist everyone to do so. Psychology's sliver-based sampling is not just a crisis for research replicability. Its implications are disturbing and – literally – far-reaching.

6 'Fitter, Happier, More Productive ...': Psychology's Exaggeration Crisis

An excess of self-esteem

As attempts to replicate their research produce a mounting series of damp squibs, you might expect that by now psychologists will have become quite cautious. Psychologists should be nervous about the way their popular paradigms contradict each other: they should surely realize that one theoretical explanation is difficult to defend alongside another that is its exact opposite. Psychologists should be equally apprehensive about their vague and imprecise approaches to measurement. They should be obsessed with equivocation: after all, those margins of error must mean something. Their statistical struggles – with *p*-values, NHST, and the like – should breed additional trepidation. Psychologists should surely react by limiting the ambition of their inferences. And each time they remember that their research captures just a thin snippet of the world's population, psychologists must feel torrents of collective embarrassment running down their spines.

Second-rate replication records, paradoxical paradigms, enigmatic measurement practices, cryptic statistics, and unconvincing sampling conventions all stand as ubiquitous reminders of why psychologists' enthusiasm should be tempered. Making evidence-based pronouncements about the human condition is a tricky business. It is best approached with humility rather than bravado.

It's true, all these reminders *should* create a climate of methodological modesty and professional shyness. But that is not what really happens. By and large, psychology is far from bashful. It is *proud* of its paradigms, claims *expertise* in measurement, sees fit to offer *its* statistical services to *others*, and largely ignores the small sizes and selective coverage of its standard samples. Psychology considers itself agile at producing authentic insights about the human psyche. It is seldom shy about its scope, impact, or achievements. Far from lacking confidence, it would appear that in some senses psychology suffers from *excessive* self-esteem.

Psychology faces many crises. But one crisis it doesn't seem to face is a crisis of confidence. Take for example the study cited in Chapter 5, which claimed that a sample of 124 North American women was sufficient to demonstrate the heretofore elusive phenomenon of human fertility signalling (Beall and Tracy, 2013). Within this study only twenty-two women were found to be wearing red clothes. Of these, only seventeen were doing so during a period of heightened fertility. Despite this, the researchers felt such numbers were sufficient to bolster a claim that women – that is, women all over the world – tend to wear red when ready for sex. In relative terms, a sample size of 124 is very small. A target group of twenty-two is tiny. And a relevant subset of seventeen is extremely meagre indeed.

In technical language, we say that small datasets lack 'statistical power'. This does not mean that small studies are unable to show real psychological effects. However, it *does* mean they can only do so if the effects in question are very conspicuous to begin with. The point is that small samples can only reveal big outcomes. Therefore, the use of tiny samples represents a fool's errand. If the sought-after effect is large enough to be detected in just a handful of participants, then it is going to be large enough to be obvious without the need of formal research.

For example, it does not take many observations to show that human beings tend to have two arms and two legs. Once you start looking at people, it becomes very obvious very quickly. On the other hand, it takes an enormous sample to show that men are a hundred times more likely than women to have protanomaly, the type of colour-blindness where people cannot see red. This is because protanomaly is relatively rare (affecting fewer than one in ten thousand women). You would need hundreds of thousands of observations to form confident projections of its prevalence. More to the point, this type of colour-blindness is so rare that, prior to hearing about it, most of us will have been unaware it even exists.

Therefore, if a sample of twenty-two red-clad women really is sufficient to show that human females indicate their sexual status through clothes choice, fertility signalling must be a glaringly obvious feature of human behaviour. It should be conspicuous to any ordinary person who pays attention to life. It should not require empirical data to demonstrate: everyone should be able to see for themselves (protanomaly notwithstanding) that women who wear red are primed for intercourse.

To argue that a particular behaviour is so subtle as to have eluded researchers for decades but, simultaneously, sufficiently *un*subtle as to be observable in a sample of two dozen people is contradictory. More specifically, to claim, without proper evidence, that women advertise their readiness for sex by wearing certain types of clothes is simply reckless. It provides an excuse to anyone who wishes to absolve themselves of culpability for sexual harassment or even assault.

The findings as published were based on tenuous evidence but nonetheless carry grave implications. That they were promulgated so widely hints at a stirring level of confidence among all those involved: the researchers, the publishing journal, and the publicists who pushed the press releases. You might conclude that their collective chutzpah extended well beyond mere confidence. Their assertions were so ahead of the data, we should calibrate their boldness accordingly. These psychologists were not simply confident. They were decidedly overconfident.

Overconfidence in science can have damaging consequences. That an unscrutinized result might reinforce negative gender stereotypes is merely a specific example. The broader difficulty is more subtle. It relates to the impact of optimism on knowledge itself. After all, scientific psychology is committed to the separation of fact from fabrication. Confidence helps us use empirical evidence to scrutinize social stereotypes. Overconfidence generates fake news that makes stereotypes go viral.

Professional loyalty can lead psychologists to focus on the strengths of psychology at the expense of thinking about its weaknesses. Methodological crises such as poor replicability, paradigmatic contradictions, sloppy measurement, and dodgy sampling can be addressed by greater vigilance. However, overconfidence is the enemy of vigilance. If psychologists develop a blindness to the faults of their field and instead acquire a habit of self-aggrandizement, then the pressure to be vigilant dissipates. The wider endeavour of progressing human knowledge becomes critically undermined.

Take a picture, it will last longer

One sign of exaggeration is excitement. One area of psychology where there is quite a lot of excitement is the study of brain imaging. Brain imaging is where the stuff of science fiction becomes the reality of scientific fact. We use machines to take pictures of human thoughts as they happen. Brain imaging allows us to commodify consciousness, to parse emotion, and to witness the birth of ideas. The evolution of the neocortex, we are told, reflects the history of the human race, and with brain imaging this narrative is now trackable in real time. Brain imaging allows psychology to concretize the abstract and to refine the phenomenon of selfhood to its elements.

Since the introduction of psychophysics in the 1800s, the idea that progress in psychology flows naturally from – and requires – technological innovation is a long-standing narrative. Brain imaging carries this mantle into the twenty-first century. According to acolytes, it is the sine qua non of psychology's cutting-edge scientific modernity. According to acolytes, that is; critics beg to differ.

It is certainly true that recent advances in brain imaging technology allow us to see more of the brain than ever before. However, contrary to widely held beliefs, the psychological implications of these developments – not to mention their cultural, philosophical, or even spiritual dimensions – are much less obvious. Clarity of a digital image is far removed from clarity of understanding. High definition is not the same as deep exposition. Pixels, in and of themselves, are not persuasive. While it is reasonable to be impressed by new technology, it does not logically follow that the images it produces will be useful.

Given the volume of media coverage and grant aid that imaging research attracts, you might expect it to have produced a torrent of breakthroughs, laying bare the inner workings of the human psyche for all to see. However, according to some commentators, the field has been far from fruitful. The shortfall is between description and explanation. Brain images will often show *what* happens, but rarely will they reveal *why* it happens.

For example, decades of mental health research have scoped out the biological aspects of depression and mood disorders. Several alterations in physiological function, including many in areas of the brain, have been found to be common in persons with these conditions. Research into the pathophysiology of depression has helped refine many treatment approaches, including drug therapies. However, none of this work explains *why* it is that some people become depressed.

Indeed, none of the documented alterations in brain function are definitively characteristic of mood disorders. Not every person with depression will exhibit disrupted function in the same brain areas. Despite the sophistication of brain imaging technology, researchers have so far failed to identify a single brain measure that is clinically useful either for diagnosing mood disorders or for evaluating outcomes of depression treatment (Savitz, Rauch, and Drevets, 2013).

Brain imaging research has many admirers, but also many detractors. One source of controversy relates to the quality of the tech. While the complexity of brain imaging is indisputable, its imprecision is frequently understated. Consider for example the most prominent imaging technique, fMRI, a method based on tracking patterns of blood flow within the brain. When the neurons that make up the brain become more or less active, they attract greater or lesser levels of blood flow. This is because neurons require oxygen, which is carried by blood, in order to function. To produce an fMRI image, a participant lies inside a scanning machine that passes magnetic signals through their head. As the haemoglobin inside blood has magnetic properties, shifts in blood flow can be detected by the scanner. The scanner can measure where these shifts occur to within one or two millimetres of accuracy. By assuming the location of the person's head, the scanner can then produce a depiction of their brain that highlights where the function is believed to have happened. The premise of the exercise is that brain activity will correspond to psychological activity. Therefore, fMRI aims to locate not just shifts in magnetism but the precise locations of specific thoughts and feelings.

While it is commonly stated that brain imaging techniques have advanced enormously during the past two decades, such an observation is inherently relative. All it means is that the technology used to be worse. To say the very least, there is much scope for further improvement.

Resolution is a particular concern. In everyday life you will be familiar with the way a digital photograph is composed of pixels, tiny spots of colour that when viewed from a distance blend together to form a detailed image. The smaller the pixels, the more detailed (and more accurate) the picture. The quality of fMRI images is evaluated in a similar way. The main difference is that because fMRI images are three-dimensional, we refer to *voxels*. In essence, voxels are the 3D equivalent of pixels, small cuboids of colour that combine to form a graphical image of the brain. fMRI is often lauded for producing images of superior resolution compared to alternative methods (such as electroencephalography, EEG, or magnetoencephalography, MEG). Nonetheless, a typical fMRI image of a brain will consist of just 228,483 voxels (Poldrack et al., 2017). That might seem like a lot of detail, but it is worth bearing in mind that the human brain contains 100,000,000,000 (i.e., 100 billion) neurons. This means that every single voxel will represent hundreds of thousands, if not millions, of neurons.

More importantly, each voxel will contain *tens of billions* of synapses, the points of contact between neurons that carry information around the brain. All theories of brain function posit that communication via synapses is integral to how the brain works. Psychology undergraduates will be familiar with the structure of synapses and the notion of synaptic transmission. Indeed, teenagers who take biology or psychology in secondary school will be aware of these principles. But these elementary processes are much too fine-grained to be studied using fMRI. In terms of explanatory power, bundling billions of synapses together in every voxel makes fMRI an extremely fuzzy technology.

Similar points can be made about the speed of fMRI. Neurons 'fire' (i.e., emit a communication signal) hundreds of times per second. However, an fMRI image of the brain takes around one second to produce. This means that every image encapsulates a blur that fuses together hundreds of separate events.

In addition, fMRI does not capture whole images in one go. Rather, every whole-brain image is a composite of several cross-sectional slice-images, each one taken at a slightly different moment. This means that when fMRI shows two brain areas to be active simultaneously, it could actually be that one was active just before the other. Furthermore, the test-retest reliability of fMRI is low, with correlations of successive measures averaging around 0.50 (Bennett and Miller, 2010). If an IQ test or personality questionnaire had test-retest reliability of 0.50, it would be considered far too unstable to be useful for research purposes.

fMRI also suffers from a poor 'signal-to-noise' ratio. This means that when some voxels 'light up' to indicate brain activity, they do so as the result of error and so should be disregarded. When researchers at Dartmouth College in the United States put a dead salmon into their fMRI scanner, they found that some areas of its brain were flagged as functional. It was as if the deceased fish had been thinking carefully about the questions the researchers had 'asked' it (Madrigal, 2009). In Linköping University in Sweden reviewers conducted three million analyses of fMRI brain images in order to compute their 'false-positive' rate: the rate at which inert voxels were highlighted as being active. At the start of the study they hoped the false-positive rate would be no greater than 5 per cent. You can imagine their shock when they discovered it was over 70 per cent (Eklund, Nichols, and Knutsson, 2016).

There are problems of human error too. For example, fMRI images are produced using software. In the three most commonly used software packages at least one significant bug has been discovered so far. The programmers who created the software made mistakes in their coding. The bug is not considered fatal, but it is believed to have distorted the results of some 40,000 studies over a fifteen-year period (Timmer, 2016).

The conventional statistical approaches used with fMRI also seem somewhat error-prone. One review of papers from the top-ranking neuroscience journals (including *Science*, *Nature*, and *Nature Neuroscience*) found that approximately half of them contained statistical errors in the way researchers compared significance levels (Nieuwenhuis, Forstmann, and Wagenmakers, 2011). Other reviewers have shown that conventional fMRI research inherently exaggerates correlational findings by failing to correct for multiple testing (Vul et al., 2009), a problem resolved decades ago in other fields of psychology.

Rampant Methodological Flexibility is also a risk. Taking account of all the parameters in a single brain scan, about which researchers are required to make judgement-based choices as to how to proceed, Poldrack et al. (2017) estimate that there are around 70,000 different ways to perform statistical analyses in any given study. As there is no single convention, it seems extremely likely that at least some researchers will be tempted to try multiple analyses until they find one that produces the results they want, even if that happens by chance.

Finally, as described in Chapter 1, the high cost of brain imaging research means that such studies typically employ tiny samples. We know that the average sample size in psychology as a whole is just too small. However, the average sample size in neuroscience is even smaller still. The typical statistical power of neuroscience studies is around 8 per cent (Button et al., 2013). This is one-tenth the power usually required for findings to be considered publishable in other sciences. It means that for every hundred possible findings, a typical neuroscience study will miss ninety-two of them simply because it is too feeble to detect a significant effect.

Nonetheless, the vast majority of published neuroscience papers report significant results. For fMRI studies specifically, the average reported effect size is substantially lower than what their small samples should be able to detect (Poldrack et al., 2017). This hints at a problem known as the 'winner's curse' (Zöllner and Pritchard, 2007). The published literature is dominated by studies that were simply fortunate enough to stumble upon unusually vivid, and thus non-representative, examples of the phenomena being investigated. In other words, the brain imaging and neuroscience literature is loaded with freakish results. Relying on it to draw conclusions about psychology is inherently hazardous.

In short, fMRI is a low-resolution, low-frame-rate, error-prone procedure that uses underpowered samples and suboptimal statistical techniques. But that's the good news. After all, each of these problems is technical in nature and so can ultimately be rectified. All we need do is wait, and eventually these problems will go away.

Illustrations are not explanations

The bad news, according to some observers, is that fMRI research (and brain imaging research in general) is hampered by a central logical limitation. The problem has been alluded to above. Imaging research provides images, not explanations. It shows us *what* things happen, not *why* they happen.

Learning that certain parts of the brain become active when certain psychological events occur provides correlational information, not causal evidence. If a particular brain region is active when a person engages in a cognition, it could be because (a) brain activity in that region causes that cognition; (b) the cognition in question causes brain activity in that region; (c) both the brain activity and the cognition are caused by something else that is not measured in the experiment; or (d) the two events happen at the same time for some reason that we just don't know. We can note that option (d) is just as plausible as all the others even though we instinctively rate the others as more satisfying. The temptation to assume we know things, and to infer causality from correlation, is very strong.

It is also worth remembering the basic premise of brain imaging research. It is assumed that brain function is linked to cognition. Such an assumption is entirely reasonable, but it has an important implication. It means that correlations between brain activity and cognitions are banal in and of themselves. It is no revelation to show that a brain activity co-occurs with a psychological event. We know this before any brain scan takes place. If the participant engages in cognition or experiences a change in emotion, some brain activity must occur somewhere. What would be revelatory would be a psychological event for which no brain activity could be found. If researchers discovered a cognition or emotion that occurred in the absence of any observable change in brain function, then this would be a huge scientific breakthrough. A blank brain image would be far more profound than one covered with nice orange blobs.

Pictures that show what brain area is active while a participant solves a logic puzzle, recalls a nostalgic memory, or is embarrassed under laboratory conditions may satisfy a technophilic curiosity. However, they do not intrinsically test any hypotheses. This is the central limitation of brain imaging research.

For example, the ventral tegmental area, a region located deep in the midbrain, becomes infused with oxygen when people look at pictures of their romantic partners (Xu et al., 2011). This may seem interesting at first. However, the finding is no more informative than learning that capillaries in the participants' faces become infused with blood at the same time. When people see their lovers, they recognize them: hence the activation of some part of the brain. And when they see their lovers, they blush: hence the infused capillaries. Rarely would we infer that blushing *causes* romantic excitement or that facial capillaries represent the source of human emotions. We do not conclude that red cheeks serve to indicate where love occurs. And yet, in brain imaging research, this is the type of conclusion that is claimed. It is believed that psychological events take place *within* whatever areas of the brain are active. Human consciousness and brain physiology are treated as one and the same construct.

It may seem that brain physiology is simply *likely* to cause human emotions in a way that facial blood flow is simply not. After all, if a person receives a facial injury, they may lose the ability to blush but retain the ability to feel love. The standard argument in neuroscience is that a brain injury would be different: if part of the brain associated with romantic emotions were damaged, then the ability to experience romantic love would be denigrated. However, this is more prediction than fact. It is a paradigm, not a point of information. Damaging the area of the brain associated with romantic love might be *expected* to denigrate the ability to feel love, but until it actually happens we will never know for sure. Brain images on their own cannot test such a prediction. They do not constitute evidence for or against the premise. Indeed, to offer the *premise* for data collection as a justification for *interpreting* the data one way and not another is an example of circular reasoning.

Simply put, it is hazardous to attribute causality to recorded brain activity. For example, for years research showed that when people consider the monetary value of consumer goods (such as when they try to weigh up a bargain), part of the brain known as the ventromedial prefrontal cortex (VPC) becomes active. This led to a widespread view that the VPC was integral to such decision-making. Eventually it became possible to test this conclusion by examining patients whose VPCs were damaged. Researchers were surprised to discover that these patients were able to perform the relevant price comparisons perfectly well (Vaidya and Fellows, 2015). The lack of apparent impairment implied that the view formed from previous brain imaging studies was simply wrong (Poldrack, 2017).

It is also unsound to assume that all causality is one way. If there is a correlation between brain activity and psychological events, it is not certain that brain activity is the cause and psychology the consequence. What appears might not be a picture of how the brain drives cognition. It might instead reflect the impact of cognition on the brain.

For example, behaviour therapy has been shown to lead to functional changes in the sensorimotor cortex (Candia et al., 2003), while the administration of cognitive behaviour therapy has been linked to activity changes in the medial orbitofrontal cortex (Schienle et al., 2009) and the rostral anterior cingulate cortex (Felmingham et al., 2007). Examples of similar effects have been noted for decades, and several have been repeatedly replicated (Miller, 2010). In one sense, it is obvious that if a treatment leads to lasting changes in the way people think, feel, or behave, then we should observe changes in their brain function too. But we should reflect on what this implies: the physiological changes (alterations in brain function) would not have occurred had psychological interventions not first been administered. In these cases, changes of thoughts, feelings, and behaviours appear to precipitate changes in the brain *rather than the other way around*.

The problem is circularity. Imaging research, such as that conducted using fMRI, can help us record the fact that many people with depression will exhibit low right-posterior activity in the brain. But a question we can then ask is:

> Are people depressed because of low right-posterior activity, or do they have low right-posterior activity because they are depressed? (Miller, 2010)

No amount of brain imaging can tell us whether depression is the cause or consequence of low right-posterior activity. In fact, the mere correlation of two organic events falls a long way short of what we need to infer causality. The intractability can be illustrated by a comparison:

> Do we gasp at the view because it is beautiful, or do we think it is beautiful because of the gasp? (Again, based on Miller, 2010)

In the end, virtually all brain imaging studies are susceptible to the same criticism. They illustrate rather than elucidate.

As with any physiological measure – such as blood pressure, adrenaline, body temperature, or the speed at which one's hair grows – recordings of brain metabolism might sometimes tell us things that introspective self-reports cannot. However, just as increased blood pressure can reflect heightened stress, heightened excitement, or heightened cholesterol, changing levels of brain activation can be deciphered in many ways. With blood pressure we can justify the research by noting that the measures are informative in and of themselves: variables associated with increased blood pressure can serve as markers of heightened cardiovascular disease risk (Hughes and Lü, 2017). But variables associated with specific brain activation do not carry the same psychosomatic implications. They can be interpreted only by referring to the researcher's prior assumptions regarding their relevance. In other words, they only make sense because the researcher says so.

Inflating the brain

Claims that pictures of the live human brain show us how the mind operates are overblown. Pictures of the live human brain show us how the live human brain operates. Implications for understanding the 'mind' remain unclear. In essence, the term 'biological basis of behaviour' is arbitrary. We should also discuss the 'behavioural basis of biology' (Miller, 2010). This is worth emphasizing, given the neuro-scientific zeitgeist that overwhelms modern psychology.

Around the world funding agencies treat biological research as more respectable than the psychological kind. Rather than resist, psychologists feel they should shoehorn as many neuroscience measures as possible into otherwise intact behavioural studies (Schwartz et al., 2016). More and more university departments are hiring neuroscience-oriented psy-chologists (Schwartz et al., 2016), and several are modifying their depart-mental titles (e.g., the 'Department of Psychological and Brain Sciences') to explicitly highlight this ethos (Lilienfeld, 2012).

Such thinking by professionals reflects the lay approach to neurosci-ence. For centuries humans have favoured materialist explanations for behaviour. They adopt a *see-it-to-believe-it* attitude that downplays the importance of inferred cognition or invisible social influence. Naïve audi-ences are disproportionately swayed by neuroscientific verbiage. When a research paper features an fMRI image, it is considered significantly more 'scientific' than the exact same paper with the image replaced by a diagram (McCabe and Castel, 2008). When bogus neuroscience jargon is included (e.g., 'Brain scans show…'), readers rate weak research reports positively (Weisberg et al., 2008). Neurobabble distracts attention from methodological flaws.

The tendency to inflate the importance of neuroscience is endemic in psychology. In this sense, neuroimaging represents the cutting edge of psychology's exaggeration crisis. Claims that scientists have found 'where trust occurs' in the brain or 'the area responsible for self-control' are unsustainable. Scaremongering about the risks of technology for children – such as warnings that screen time elicits the same brain activity as narcotic abuse (Lin et al., 2012) or that social networks cause cortical damage that in turn precipitates autism (Greenfield, 2014) – are unrelated to anything actual research could ever have revealed (Bishop, 2014). Assertions that cell phone use impairs 'the part of the brain responsible for driving' (Miller, 2010) or that specially programmed puzzle apps can prevent age-related cognitive decline (Walters, 2016) amount to little more than public miseducation.

Even the mainstream practice of depicting mental illnesses as brain disorders can obscure rather than reveal. Conditions such as depression and schizophrenia undoubtedly have physiological aspects. Differential patterns of brain function are certainly relevant to consider. However, as outlined above, directions of causality in such associations remain unknown, despite thousands of brain imaging studies. Indeed, a focus on biological causality may distract from more important factors.

For example, one of the strongest statistical predictors of psychiatric ill-health is poverty: mental illness is far more prevalent among the poor than the rich. This is rarely acknowledged in psychiatry journals, which remain saturated with reports invoking neurochemical, anatomical, genetic, and pharmacological frameworks (Kagan, 2013). Likewise, income inequality and social exclusion almost single-handedly account for aggressive behaviour, yet researchers appear permanently embroiled in a hunt for the (thus far elusive) 'gene for aggression' (Thomas and Pope, 2013). And drug addiction is frequently cast as a 'disease of the brain' on the basis that it leads to lasting changes in brain function, even though aerobic exercise also leads to similar changes without there being a wish to depict jogging as a brain phenomenon (Miller, 2010).

Brain imaging research exemplifies each of the crises outlined in this book. The exploratory (as opposed to predictive) tenor of brain imaging, not to mention the low reliability of fMRI measures, contributes to poor replicability (Boekel et al., 2015). As such, neuroscience is deeply embroiled in psychology's **replication crisis**. The **paradigmatic crisis** is highlighted by the muddled way neuroscience is both absorbed and ring-fenced within psychology. When university units rename themselves as Departments of Psychological and Brain Sciences, are they implying that the two named specialisms are distinct? Research employing technology to record brain activity has been part of psychology since the late nineteenth century. Separating the terms when naming a department amounts to logical confusion. As pointed out by Schwartz et al. (2016), the revised nomenclature is akin to titles such as the 'Department of Psychology and Personality Research'.

The fact that technical limitations of brain imaging are so frequently overlooked exemplifies psychology's **measurement crisis**. Brain images are seldom as precise as they seem. Error rates are high, while test-retest reliability is low. Psychology's **statistical crisis** is also apparent. As described above, statistical methods that are customary in brain imaging are considered weak by the standards of other fields. And the prevalent use of very small samples, resulting in minuscule statistical power, places brain imaging research at the heart of psychology's **sampling crisis**.

The fact that such studies are destined to overlook the vast majority of possible true findings, but contribute a disproportionate number of false positives, brings together the worst of both worlds. Add in the inherently correlational nature of imaging data, as well as the rather dubious implication that brain activity is synonymous with psychological experience, and we are well on our way to a crisis of exaggeration.

But exaggeration is not exclusive to biological reductionism, nor does it require technophilia. Exaggeration is just as likely to emerge from psychologists who are sceptical of physiology and gadgets. Researchers hostile towards techno/bio worldviews are just as prone to overselling outcomes and overlooking faults. They are just as liable to fall into the exaggeration trap. It is to one such example that we now turn.

Chronic fatigue syndrome: Psychologists to the rescue

To many non-psychologists, the provision of therapy is the raison d'être of psychology. However, for as long as they have existed psychotherapies have been controversial. Many psychologists find criticism of psychotherapy very hurtful, viewing it as an attack on their professional integrity. But if vulnerable clients could benefit more from non-psychological treatments (such as drugs or surgeries), then it is ethically dubious for psychologists to insist that talk-based or behavioural interventions should be used instead. Therefore, it is a matter of public interest that psychologists avoid exaggeration when championing their own forms of treatment. Psychologists should be careful not to defend psychotherapies without robust empirical evidence. They should especially not do so in *defiance* of such evidence.

One high-profile psychotherapeutic controversy has concerned the use of psychological interventions for myalgic encephalomyelitis, a condition alternatively referred to as chronic fatigue syndrome (ME/CFS). According to the US National Academy of Medicine, ME/CFS is a serious debilitating physical illness characterized by prolonged and significant decrease in function, on-going fatigue, post-exertional malaise, and many other difficulties, including pain (Institute of Medicine, 2015). Patients diagnosed with ME/CFS experience sudden onset followed by years of debilitation. They are unable to hold down jobs or meet family responsibilities. Around one in every three hundred people will experience ME/CFS, equating to 250,000 cases in the UK alone. All will require some form of medical assistance.

The cause of this condition is unclear. Several studies have pointed to abnormalities in patients' brain, muscle, and immune function (Goudsmit and Howes, 2017). Recent research has found that patients exhibit extensive cell-functioning impairment arising from quirks in DNA (Marshall-Gradisnik et al., 2016). They also show increases in several immune system proteins (Montoya et al., 2017). Such findings corroborate claims that ME/CFS results from immune hyperactivity. The condition often arises after a physical trauma, such as an infection or accident, and it appears that in some people immune activity fails to shut down properly once the trauma has passed. Nonetheless, while research continues, the precise cause is unknown.

The perception that ME/CFS is a mystery illness has led some healthcare professionals to conclude it is not a *real* illness. Doubt has been evident for decades. In a landmark *British Medical Journal* paper from 1970, reviewers concluded that ME/CFS was a diagnostic illusion, produced by a combination of mass hysteria in patients and gullibility among doctors (McEvedy and Beard, 1970). This verdict is mentioned in nearly every historical account of ME/CFS. However, with the benefit of historical hindsight, the *BMJ* conclusions look weak. Essentially, the authors presented two main arguments. Firstly, they claimed the condition must be anecdotal because its most common symptoms were measured using self-reports. It seems they felt that patient testimonies are intrinsically untrustworthy. Secondly, they claimed that ME/CFS was likely to result from mass hysteria because their data showed it affects women more than men. Their unstated logic was that female brains are addled by irrationality; therefore, when women complain of an illness it is likely to not exist.

Despite its limitations, the *BMJ* paper helped establish a view that ME/CFS is psychological rather than physical. Accordingly, while the major British and American research agencies have pumped £10 million into studies of ME/CFS, most of this funding has supported studies of psychological theories and behavioural interventions. By far the biggest has been the British PACE Trial (short for 'Pacing, graded Activity, and Cognitive behaviour therapy: a randomised Evaluation'). This trial has attracted almost £5 million worth of public funding (McPhee, 2017), making it one of the most expensive studies of psychological therapies ever conducted in a healthcare setting.

The PACE Trial examined psychological treatments for ME/CFS. It focused on cognitive behaviour therapy (CBT) and graded exercise therapy (GET), interventions favoured by practitioners who feel that ME/CFS is a psychological problem. The investigators studied 641 patients for one year. Each patient was allocated to one of four treatment groups: CBT; GET; 'adaptive pacing' control (in which patients were advised on how to reduce activities in proportion to energy levels in order to avoid fatigue); and no-treatment control (in which patients received no treatment other than standard medical care).

The researchers reported that around 60 per cent of participants in the CBT and GET groups exhibited substantial 'improvement' in their condition, compared to around 45 per cent of patients in the control groups (White et al., 2011). More importantly, they reported that 22 per cent of the CBT/GET patients actually 'recovered' from ME/CFS, compared to just 7 per cent of patients in the control groups (White et al., 2013). Simply put, the authors claimed to have found a cure for ME/CFS. That cure was psychotherapy.

Taken at face value, the PACE Trial appeared to have made two major breakthroughs. Firstly, the results suggested that CBT and GET were highly effective treatments for ME/CFS. These therapies do not involve costly drugs, surgeries, or periods of hospitalization. Neither do they require time-investment by doctors, whose salaries are high and training expensive. Paying counsellors and health psychologists to provide the relevant therapies is much cheaper than paying disability benefits, over a lifetime, to debilitated patients.

Thus, not only were the interventions medically effective, they were also cost-effective: using them would save service-providers, insurance companies, and governments millions of pounds. As a direct consequence of the PACE Trial, CBT and GET were declared 'evidence-based' treatments for ME/CFS by the UK's National Institute for Health and Care Excellence, the body charged with issuing guidelines for clinical practice in Britain's health service.

The second breakthrough was to corroborate a psychological basis for ME/CFS. If a disease can be dealt with by psychological therapy, it seemed likely to be a psychological disease. Thus, in the multidisciplinary drama surrounding this strange condition, psychologists had come up with the answer. Not only could they cure this mystery illness, they could also explain it. The PACE Trial highlighted how valuable psychologists can be.

But not everyone was convinced. Many ME/CFS specialists considered the results of the PACE Trial to be simply unbelievable. Few doubted that psychological therapies could be useful adjuncts to help patients cope better with physical disease. However, it seemed incredible to claim that such therapies could elicit authentic 'recovery' from an illness such as ME/CFS. Patient groups complained that psychological interventions had no effect on their core symptoms. They noted that GET was known to be particularly unhelpful, with many patients finding it materially counterproductive (Kindlon, 2017). And while clarity had not yet been achieved, it seemed only a matter of time before biomedical research would uncover the condition's physiological cause, thereby casting doubt on the idea there could be an effective psychological cure.

Controversy hit fever pitch when patient groups and sceptical researchers asked to see the original data. For researchers a willingness to share data is both an ethical obligation and a condition of publication. However, the PACE Trial investigators declined all data-sharing requests for several years. Ultimately, after the matter went to court judges decided that the data should be made available to the public, noting that the work was funded – and thus owned – by taxpayers. Half-a-decade after the original study report was published, the raw data were finally released. When independent reviewers reanalysed the figures, they did not like what they saw.

The PACE Trial as a microcosm of psychology's crises

You will recall that psychology's fundamental **replication crisis** is propelled by a tradition of Rampant Methodological Flexibility. Psychologists are free to craft bespoke research designs and analytic approaches. They can tweak any and all details as they go along. They can analyse and re-analyse data using many different approaches and then continue to tinker until they find a result they are happy with. This opens the door to unconscious bias. If findings contradict their preferences or assumptions, researchers can simply look for alternative analytic strategies. If findings *support* their preferences or assumptions, they can leave well enough alone. Whenever results are eventually published, readers will be totally in the dark as to what arbitrary methodological choices the researchers have made. With so much room to customize, it is little wonder that most psychology studies are nearly impossible to replicate.

When their datasets were independently scrutinized, it became apparent that the PACE Trial authors had made several arbitrary choices. One related to the way they described 'improvement' and 'recovery'. When first designing the PACE Trial, the authors drafted 'improvement' and 'recovery' guidelines that were based on a number of measures. One was an instrument known as the Short-Form Health Survey (SF36; Ware and Sherbourne, 1992). To be deemed as having 'improved' (or 'recovered') after treatment, patients needed to score 85 or above on this questionnaire. At the end of the trial, it became apparent that only 20 per cent of patients in the CBT/GET groups had met this standard (Wilshire et al., 2017). This made CBT and GET look rather disappointing: four out of five patients saw no benefit from the interventions.

The PACE Trial authors came up with a novel way to handle their disappointment. They changed the criteria. They decided that SF36 scores of 85 would no longer be required to indicate 'improvement' or 'recovery'. Instead scores of 60 or better would be sufficient. All of a sudden, the 'improvement' rate for CBT and GET grew from 20 per cent to a rather more impressive 60 per cent. In other words, *three times* as many patients were now deemed to have 'improved'. Obviously, rates of 'recovery' were also boosted. The authors were now able to report that 22 per cent of CBT/GET patients 'recovered' from ME/CFS (White et al., 2013), a condition previously understood to be incorrigibly resistant to treatment. In reality, only 7 per cent of the patients met the original definition of 'recovery' that the PACE Trial was designed to assess (Wilshire et al., 2017), no better than the rate of *spontaneous* recovery seen in the trial's own control groups.

Moving the goalposts like this seems obviously controversial. But in terms of replicability, the PACE Trial's most significant weakness was its core research design. There were two big shortcomings: (a) the trial was not 'blinded', even though it relied on precisely the kind of data for which blinding is considered essential; and (b) the trial employed non-equivalent control conditions that served to magnify the relative superiority of the target therapies. Either of these lapses on its own would be damaging. Both in combination are ruinous.

'Blinding' is a research practice used to reduce expectancy bias in therapy studies. In a blinded study, participants will not be told the purpose of the treatment they are to receive. When participants have such information, it can affect how they evaluate their experiences. If they believe they are getting a 'strong' therapy, they may interpret ambiguous symptom changes as improvements; if they believe they are part of a control group, they may perceive the same changes to be trivial. Blinding ensures that participants do not know whether the treatment they receive is special, standard, or spurious. They will, thus, be 'blind' to their place within the overall research endeavour. Blinding is considered useful in all therapy studies, but especially necessary where outcomes are measured using self-report. This is because, by definition, self-reports are bound up in exactly the kind of subjective impressions that produce expectancy bias in the first place.

Given that the PACE Trial's key outcomes were derived using self-report, you might anticipate that the study featured extensive, if not ingenious, attempts at blinding. However, this was simply not the case. In fact, far from being blinded, participants in the PACE Trial had their attention persistently drawn to their place in the research effort. They were provided with detailed promotional material extolling the benefits of CBT and GET. They were sent documents in the post that described CBT and GET in glowing terms. Somewhat pre-emptively, the bumf even referred to the therapies as 'evidence-based' (Geraghty, 2017).

Participants in the CBT and GET groups were also invited to attend five times as many appointments as participants in the control groups, creating a non-equivalence problem. Given that patients usually prefer to be seen than to be neglected, such variation in exposure to service-providers could alone account for differences in self-reported outcomes (Vink, 2017). In short, rather than taking steps to *avoid* the threat of participant expectancy bias, the PACE Trial authors could hardly have done a better job had their intention been to actively *cultivate* such an effect (Edwards, 2017).

The PACE Trial also exemplifies psychology's **paradigmatic crisis**. Conceptually, the trial was premised on the view that ME/CFS results from cognitive factors and not biological ones. This reflects the investigators' paradigmatic preferences. The PACE Trial team included a number of psychologists who had previously publicly promoted the use of CBT and GET as ME/CFS treatments (Lubet, 2017). Such allegiance to a therapeutic paradigm is challenging in a research context. Researchers might be reluctant to acknowledge negative findings that would discredit their prior writings or positions.

With ME/CFS, the allegiance to a psychogenic explanation, in spite of mounting evidence of physiological causality, might also explain why the PACE Trial authors occasionally accuse patient organizations of 'attacking science' (Edwards, 2017) simply for raising criticisms of the trial's methods and procedures. It is as though non-cognitive counter-explanations are seen as dogmatically unacceptable and, thus, automatically unscientific.

Psychology's **measurement crisis** is writ large in the way the PACE Trial assessed 'improvement' and 'recovery'. We have already discussed how movable thresholds mess up therapy trials. But in the PACE Trial the way in which the authors revised the recovery criteria was especially peculiar. The new benchmark for being deemed to have 'recovered' – an SF36 score of 60 – was *lower* than that required to be eligible to participate in the first place. To qualify for the PACE Trial, patients had to have SF36 scores of 65 or less. This meant that some participants exceeded the modified 'recovery' threshold *before the trial even began*. They were sick enough to take part but simultaneously well enough to be classified as having 'recovered'. Around one in every eight participants was in this strange situation. It should go without saying that for this subgroup the fact that improvement pre-dated therapy debunks any claim that it resulted from it. Indeed, some participants' scores *declined* and yet they were still classified as having 'improved' (Vink, 2017; Wilshire et al., 2017).

In summary, the PACE Trial's assessment of 'improvement' and 'recovery' was notably poor. The measurement lacked construct validity. It lacked reliability. It was so fluid as to be bidirectional: whether your score went up or down *didn't necessarily matter*. It is obvious that the value of a therapy trial depends on how well the outcome of the therapy is assessed. That one of the biggest ever such trials would approach the task in this lax manner hints at a very deep crisis indeed.

The PACE Trial also offers glimpses of psychology's **statistical crisis**. Like much psychology research, the authors placed undue emphasis on statistical significance at the expense of effect size. By focusing on the *proportion* of participants who had (purportedly) recovered, attention was drawn away from the *amount* of recovery that each patient had (or had not) exhibited.

The reality is that, even for those patients who were supposed to have benefitted, the relative impact of CBT and GET seemed very modest. Even were we to accept the PACE Trial's own claim that 60 per cent of patients who received CBT or GET had 'improved', we can note that the average amount of improvement in each case was wholly underwhelming. People's conditions improved but mostly by just a tiny amount: when the released datasets were independently scrutinized, it was found that, in statistical terms, the effect sizes for the PACE Trial were mostly 'small' and 'medium' (Stouten, 2017). This is far removed from the original impression that the trial had found a miracle cure for a mystery illness.

Finally, and perhaps unsurprisingly, the PACE Trial was no exception when it came to psychology's endemic **sampling crisis**. It is something of a geographical necessity that the trial focused on patients from the United Kingdom alone. But it also attracted criticism for the way in which it sampled these patients. There are several methods for formally diagnosing the ME/CFS. The PACE Trial followed the 'Oxford criteria' (Sharpe et al., 1991), an approach that emphasizes chronic disabling fatigue as the condition's primary symptom (Vink, 2017). The Oxford criteria are certainly widely employed, but they are by no means universally accepted. Many specialists feel they are too loose.

For example, the 'Canadian consensus criteria' (Jason et al., 2010) specifically exclude diagnosis in cases where participants exhibit mental illness. In the PACE Trial, 47 per cent of participants had comorbid mood or anxiety disorders. This means that half of the PACE Trial's participants would not have been diagnosed with ME/CFS had they lived in Canada. Other specialists recommend an approach known as the 'London criteria' (National Task Force, 1994). Again, around half of the PACE Trial's participants would have been excluded from an ME/CFS diagnosis had those criteria been followed (Vink, 2017).

The PACE Trial is not unique in offering an optimistic view of psychological intervention. It reflects a general view seen in therapeutics in many areas, including biomedical and rehabilitative treatment. People generally *want* therapies to work. Both specialists and non-specialists seem to overestimate the benefits of therapies while underappreciating their potential for harm (Hoffmann and Del Mar, 2015).

Psychological interventions are notoriously difficult to research for several reasons. Research into psychotherapies has a long history of producing findings that are at once underwhelming but nonetheless oversold (Dragioti et al., 2017). As one of the largest ever studies of psychotherapeutic interventions in a healthcare context, it is possibly inevitable that the PACE Trial would exhibit many of the problems that plague psychology as a whole. Given its impact outside of psychology – in raising doubts about the evidence for clinical care guidelines that affect the lives of hundreds of thousands of people – the controversies surrounding the PACE Trial can be seen as emblematic of the real-world problems caused by psychology's many crises.

The inevitability of inflation

We have seen that psychologists sell their findings with pride, even when the platform underlying their claims is somewhat rickety. The portrayal of brain imaging as cutting-edge, without reference to its elementary shortcomings, is but one example of how biologically reductionist paradigms are afforded questionable pre-eminence in psychology. But the exaggeration crisis is not confined to psychologists who become awestruck by gadgetry or deferential towards laboratories. It is also apparent in the parts of psychology that promote talk therapy and human interaction as intrinsically virtuous endeavours. That the PACE Trial continues to be so doggedly defended, despite a litany of damaging critiques, shows us how psychologists can retain an unswerving allegiance to their own ideas. It reflects a general tendency for psychologists to promote psychological interventions over alternatives (e.g., over drug treatments) to an extent that seems out of sync with the balance of available evidence.

What is it that encourages psychology's exaggeration crisis? What causes psychologists to pump up their claims about psychology? Where does the hype come from? Well, some of it undoubtedly results from old-fashioned attribution bias. Psychologists have long pointed out that people generally tend to interpret ambiguity in self-flattering ways. Psychologists are people; presumably, therefore, they too attribute positive aspects of their work to merit, and negative aspects to chance. The net result is a genuine belief that brain images are profound, that one's own therapy is outstanding, and that psychology's research findings are more robust than is actually the case. In short, most of the time, the outcome of attributions is exaggeration.

Some inflationary influences are external. One problem is the perennial competition for public attention. By and large, people do not welcome complexity. They live every day preoccupied by their own lives and interests. Insofar as they consume information, they want it to be easy to digest. This is why simple, short, and attention-grabbing information is more popular than information that is abstract, complex, or epistemologically challenging. Such circumstances accelerate the spread of misinformation, quick fixes, and bad ideas. Peer-reviewed scientific research reports must compete with mass traffic websites, ad-laden news channels, and ordinary hearsay. Science, which endures a laboured process of validation, is required to coexist with pseudoscience, which is unfiltered and immediately available (Hughes, 2016).

Psychology bears the additional pressure of mass public fascination. Families rarely talk about subatomic physics at the dinner table, but they most likely discuss human feelings and behaviours every day. The popularity of its subject matter ensures that psychology is always in the spotlight. As a result, psychologists are under constant pressure to explain their work in the simplest of terms. One of the easier ways to simplify is to exaggerate: cut out the caveats and emphasize the impact. Therefore, it should be little surprise if many psychologists become accustomed to inflating the importance of their field.

More broadly, this preference for simplicity has biased human culture towards materialism. People trust what they can see and touch more than what they must visualize or think about (Kagan, 2013). This leads to a tendency to prioritize concrete depictions over abstract explanations. Once again, psychology is under pressure to simplify its message. Bald assertions that 'The therapy worked!' are much more welcome than warnings about confounding variables, margins of error, or skewed samples. A materialist bias also makes people receptive to the idea that the explanation of complex cognition can be reduced to a diagram of brain parts. When audiences crave concreteness, psychologists will find that exaggeration helps them to be heard.

Psychology must also compete in various institutional and policy-related domains. Professional organizations that represent psychology, and member-psychologists who in turn represent their professions, will often be called upon to 'make the case' for psychology. Universities, research institutions, and policy agencies all have to manage a multiplicity of scientific and academic subject areas. As available resources are finite, the success or otherwise of a subject area will depend on the share of resources made available to it. For example, in universities, psychology departments must convince administrators of their need for financial support for both teaching and research activities. Given that virtually all arenas in which psychology operates are competitive in this way, there is a pressure on psychologists to emphasize their field's positive contributions while playing down its more embarrassing aspects.

Applied psychologists are engaged in equivalent struggles. For example, in health services clinical psychologists must present arguments for the utility of clinical psychology, lest their work be farmed out to other health practitioners. In effect, psychologists are constantly promoting their own livelihoods. That they might develop a habit of overlooking their own faults seems entirely understandable.

It is also worth considering the channels through which psychology knowledge is conveyed. As with other sciences, psychology research is considered credible only if it appears in peer-reviewed academic outlets such as scientific journals. These journals are published by a variety of organizations, some of which are funded by governments in the public interest, others of which are run as commercial activities by corporate entities. The traditional pre-publication quality-control approach is to ensure that every paper is reviewed by independent and appropriately qualified peer scientists. Post-publication, the worth of a work is inferred from how frequently its findings are cited by other scientists in later reports. The relative quality of different journals is evaluated by studying each one's overall citation data over a given period. Ultimately, the research community then weighs up the state of evidence on a given topic by gathering all the relevant research reports and considering their relative merits in an objective way. The overall system is assumed to produce reliable scientific knowledge on the basis that all its features operate as they are superficially designed to do.

However, as with all human endeavours, this system has frailties. In Chapter 1 we noted how journals only ever carry a fraction of the research that has been conducted in the world. Therefore, it will always be impossible to know whether work that appears in print is truly representative of what scientists have been doing with their time. We know that journals reflect the materialist bias seen in culture at large: they favour the publication of concrete results (that is to say, statistically significant hypothesis tests) over indecisive ones. In other words, we can safely say that journal contents are *not* representative of what scientists do with their time. In fact, the dominant publication practices encourage, rather than suppress, exaggeration.

Professionally, scientists and academics are judged on how many papers they publish. The more publications, the more promotions. Therefore, as is often quite evident from their online CVs and personal webpages, there is a clear imperative for research psychologists to blow their own trumpets to high heaven and back. You could even say that those who *fail* to exaggerate are behaving irrationally in wilfully choosing to undermine their own self-interest.

Competition for attention, a cultural bias towards materialism, interdisciplinary hunger games, journal practices that promote hysteria, an effective requirement for professional self-aggrandizement, and old-fashioned attribution bias all combine to produce psychology's exaggeration crisis.

Problems of poor measurement, statistics, sampling, and the like can be tackled in various practical ways. Attempting to solve the problem of exaggeration will require a change in mindset and a sudden bout of heretofore scarce humility. In this way, the problem of exaggeration draws attention to the apparent intractability of psychology's crises. In our final chapter we will examine the most pernicious crisis of all – the crisis of *being in crisis* – and what psychology might usefully do to pull itself out of the abyss.

CHAPTER 7

From Crisis to Confidence: Dealing with Psychology's Self-Inflicted Crises

The crisis of being in crisis

Some people claim that with every passing decade, the world just gets worse and worse. Our trajectory is ever towards turmoil. Social chaos surrounds us, war and suffering are commonplace, upheavals are unexceptional. As the world becomes increasingly interconnected, life gets exponentially more stressful. Everything is always changing. Constantly we are called upon to catch up; staying still has become stigmatized. Getting through each day requires flexibility and alertness, pragmatism and luck. Human affairs are shaped by expedience as much as by empathy. Life is noisy and fitful. Crisis, it seems, is permanently in the offing.

A common claim is that things were easier in the past. Life was less controversial. You could let children play on the street, leave your house unlocked, walk alone at night, not be killed by terrorists. But many scholars argue the opposite: that over time our world has become safer, fairer, and more peaceful. True, our news media report more bloodshed and brutality nowadays, but that says more about contemporary journalism than it does about the planet we inhabit. In reality, the average human has never had it so good. They enjoy more freedom, greater health, less oppression. Enlightenment has enhanced their social order; technology keeps them comfortable and safe. They are less likely to die prematurely or violently than at any previous time in history.

As always, truth lies somewhere between the extremes. In many senses we are indeed luckier to be alive now than in the past. But not all of us. Around one-fifth of the human race lives under threat of political violence (Etchells, 2015). And while wars are declining in number, they are becoming more deadly. In 2014 the number of people killed by war exceeded 100,000 for the first time in a quarter of a century (Gates et al., 2016). By

2016 it had risen to 160,000 (IISS, 2017). True, relative to global population growth, such casualties comprise shrinking percentages of overall humanity. But if human suffering is de facto a bad thing, shouldn't volume of death, not proportion, be considered more important?

Judging chaos is not easy. Researchers can be optimistic or glum. Studies suggests that most human beings hold positive hopes for the future (Jefferson, Bortolotti, and Kuzmanovic, 2017) but equally that many are prone to 'declinism', the enduring belief that things will never be as good as they were before (Etchells, 2015). Such attitudes affect researchers too. Historical wars seem worse if you focus on battlefields, modern ones seem worse if you focus on civilian casualties (Lewis and Lewis, in press). In evaluating human warfare (or welfare), optimists and declinists will draw different conclusions from similar data.

Equivalent ambiguities arise when evaluating psychology's welfare (or warfare). Declaring a crisis relies on judgement, on interpretational choices, on *attitude*. Psychology is a human activity, perhaps the most human activity of all: pondering and probing one's own mind and identity and those of others. It is not just a science, it is a social science. And not just a social science, but a *very* social science. The dynamics of psychology resemble those of human affairs in general. If the world is in permanent crisis, why should psychology be any different?

Life comes at you fast. Crisis is part of the package.

Crisis as endemic in psychology

Psychologists have been talking about crisis for a very long time. In the late nineteenth century, with the field in infancy, an obscure Swiss researcher was already bemoaning '*die Krisis in der Psychologie*' (Willy, 1897). In the early twentieth, an obscure French one declared '*la crise de la psychologie expérimentale*' (Kostyleff, 1911). The disquiet eventually went viral: in the 1920s you could read about psychology's crises in mainstream European newspapers (Sturm and Mülberger, 2012). Crisis talk ebbed and flowed for decades, sometimes intensifying, other times waning, but never completely going away.

Some crises are philosophical, others are more practical. An enduring problem for clinical psychology has been the medicalizing of mental healthcare, a symptom of the dubious symbiosis that exists between Big

Pharma and professional psychiatry. In modern pharmaceutics commercial concerns can consume more energy than empirical ones. Much of the industry's innovation relates to branding (as when 'tranquilizers' get renamed to 'antipsychotics') rather than the development of new drugs that exceed the efficacy of old ones (James, 2016). The entanglement of private drug production with public healthcare skews the focus and interpretation of research, discourages new science on non-drug treatments, and limits investment in disease prevention. It is an all-round crisis for psychology.

Sometimes psychologists face crises of their own making. In 2015 the American Psychological Association was embroiled in a scandal concerning torture. An independent commission claimed that the APA had arbitrarily diluted its ethical guidelines to provide legal cover for torture-style interrogations at Guantánamo Bay, the controversial US military prison where detainees are held without trial (Risen, 2014). According to investigators, the APA wished to curry favour with the military so that it would hire more psychologists. Rival professions (psychiatrists and medical doctors) had publicly condemned the interrogations, so the APA tweaked its ethics code to ensure that psychologists would be seen to have no such scruples. These findings have been disputed (Engber, 2017), but not before several APA officials either resigned or were fired (Ackerman, 2015). The crisis undoubtedly damaged the reputation of American psychology. It never looks good when a professional body seems willing to overlook torture.

Politically, scientifically, or philosophically, psychology seems always on the verge of cataclysm. It is forever facing existential decisions and navigating crucial turning points. But whether the decisions are ever made or the corners ever turned is an open question. The various crises discussed throughout this book – those relating to replication, paradigms, measurement, statistics, sampling, and exaggeration – all have lengthy histories. If anything, the ability of psychologists to ignore crisis talk and carry on regardless indicates its own existential crisis: a crisis of incorrigibility.

The rigidity of crises

In some respects, intermittent crises are the heartbeats that drive human history. They are a recurring motif in the story of civilization. Traditional history is often portrayed as a series of stormy events in which individual societies are dramatically reshaped into new structures, systems, and stratifications. Social values and cultural norms also shift, sometimes as the result of political upheaval, sometimes as its cause. This turbulent flow is not restricted to politics and culture; science is part of it too.

Philosophers have observed this for a very long time. Heidegger (1927) advised that 'the level which a science has reached is determined by how far it is capable of a crisis in its fundamental concepts'. Popper (1932) wrote of crisis as being 'the normal state of a highly developed rational science'. Kuhn (1962) famously argued that scientific progress is punctuated by revolutions. These days commentators laud the merits of 'disruptive innovation', where new ideas and technologies reorder the norms of daily life (Christensen et al., 2006). If psychology is indeed permanently in crisis, then maybe this indicates forward momentum. Perhaps a state of crisis is the desirable endpoint for serious science.

Finding consolation in ambient catastrophe might be taking optimism just a little too far. After all, tragedies and disasters generate a sense of crisis without there being an obvious upside. Maybe psychology's crises are what they appear to be: symptoms of serious malaise.

Dysfunctional psychology may result, in part, from scientific illiteracy. Survey after survey reveals that most people are poorly equipped to interpret the results of, well, surveys. In fact, we are naturally ungifted at understanding *most* forms of empirical research. This frequently leads to strange views and ideas. In developed countries more citizens believe in ghosts (YouGov, 2012) than in the theory of evolution (Gallup, 2017). Healing activities such as past-life regression, psychic surgery, homoeopathy, bioenergy therapy, reiki, radionics, reflexology, rebirthing, cupping, chiropractic, crystal therapy, craniosacral therapy, colonic irrigation, human urine therapy, and Hopi Indian ear candling attract crowds of customers despite being scientifically debunked. Few people are born scientifically literate. Psychologists are no exception.

Many fully trained psychologists harbour attitudes to science that are quite lukewarm. They feel that methodological debates are of peripheral relevance to their work (Hughes, 2008). Their personal identities are bound up in using psychology as a clinical tool, rather than in conducting high-quality research studies. Scientific illiteracy makes them poor at detecting, or discouraging, error.

Even a fair portion of academic psychologists are relatively under-skilled in scientific methods. As pointed out in Chapter 4, only a minority of staff in our university psychology departments feel comfortable with anything other than basic statistical analysis. These specialists are seen by their fellow academics as occupiers of an obscure curricular niche, colleagues with skills that are unique rather than universal, gurus down the corridor who can be consulted when one's own research study begins to unravel (but not before then).

Many psychology lecturers teach subject-area courses by referring to prior research as though all that matters is the name of the author and the year the paper was published. Students are not often encouraged to ask whether the studies so cited were rigorously designed or whether their findings were ever replicated. They are invited to name-drop citations to pad out their essays, rather than build a case for a conclusion using the best empirical evidence. One of the more sobering outcomes of the replication crisis has been the revelation that many of the so-called classic studies featured in undergraduate textbooks describe findings that cannot be reproduced.

One reason that psychology's crises can seem incorrigible is that they are recycled from generation to generation through an education model that is itself hampered by a tendency to avoid difficult questions. In short, the majority of psychologists – including the majority of those employed to educate the oncoming generations – are cautious around methodological controversy. They adhere to common custom and practice, repeating what they themselves have been taught, perpetuating a copy-and-paste consensus. Old methods identified as suspect in the 1960s are wheeled out time and again as if fit for purpose in the twenty-first century. The fact that the majority of those who teach psychology see no problem with the status quo, and so say nothing about it, does not indicate that their discipline is healthy. If anything, it implies the presence of groupthink. One might even consider it an instance of mass delusion.

Who benefits from crisis denial?

As noted in Chapter 1, some psychologists claim there are no crises in psychology. According to them, the field is in rude health. They are not perturbed by psychology's incompatible paradigms. They believe psychology's measurements are immaculate. Yes, there are debates about statistical approaches. But a crisis? Not really. Psychology's sliver-based sampling causes them no concern. They dismiss the idea that psychologists systematically exaggerate their offerings. And most of all, they feel the replicability of psychology's research is just fine, thank you.

It is not hard to see why psychologists would think this way. Sheer self-interest will be a factor for some: those whose own research has been directly debunked will find it particularly hard to muster true objectivity. Even some high-profile psychologists are now being challenged to consider whether 'everything they've been doing for years is all a mistake' (Gelman, 2016). Few would willingly go that far. It is understandable if instead they cling to circuitous rationalizations, choosing not to admit to folly.

Implicated individuals are not the only crisis-deniers in psychology. In public debates at conferences, on message boards, or in the glare of social media, there will inevitably emerge a view that all this talk of crisis is inflated, if not fantastical. Some scepticism will stem from solidarity with those psychologists whose non-replicable studies have been publicly shamed. However, many will feel a generic type of defensiveness. It is not necessary to have particular cases in mind when defending psychology's honour.

This may be accentuated by a feeling that psychology's crises are generational. Crisis narratives usually contend that the old ways of doing things are flawed. Often the subtext appears to be that the old *people* are flawed too (Spellman, 2015). As with most professional groups, the hierarchies and power structures in professional academia have evolved very slowly and are not known for spontaneously inviting change upon themselves.

The resulting paranoia can lead to hasty argumentation. For example, defenders of psychology often claim its rate of false positives must be below 5 per cent, because that is the threshold used to determine statistical significance (i.e., $p < 0.05$). This very commonly cited reasoning is logically flawed (Pashler and Harris, 2012). Using p-value thresholds to estimate false-positive rates would make sense only if *all* statistical tests ever performed from *all* studies ever conducted were in the public domain. In reality, only a subset of analyses from a subset of studies are ever reported. The available literature is the tip of the iceberg; unpublished research, languishing in researchers' file drawers, is what lies below the waterline. The norm that significance is more publishable than non-significance skews the content of journals accordingly. Therefore, when estimating psychology's error rate, *p*-value thresholds are irrelevant.

Another common defence is that, over time, scientific vigilance will prune erroneous findings from the literature. Quite where these prunings are supposed to go is unclear. Around 2.5 million journal articles are published each year (Ware and Mabe, 2015). However, fewer than six hundred of these are retracted (McCook, 2016). If it is indeed true that 'most published research findings are false' (Ioannidis, 2005), then a lot more retractions will be required before the desired level of pruning is achieved.

More to the point, erroneous findings can only be pruned if they are discovered to be erroneous. In the absence of a replication culture, fallacious findings will linger in the literature forever unchallenged. There is little to suggest that scientists are good at rooting out dodgy research. While egregious fraud is rightly frowned upon, sloppier self-serving inefficiencies are usually overlooked (John, Loewenstein, and Prelec, 2012).

Who favours crisis talk?

While some observers deny psychology's crises because of unconscious (or conscious) bias, others exaggerate them for similar reasons. The idea that psychology might not be all it's cracked up to be can deliver a certain degree of comfort. Some groups see benefits in crisis talk.

For example, social conservatives – who as a point of principle wish to preserve the traditional cultural arrangements of society – often appear particularly sceptical of psychology and only too pleased to learn when its methods might be suspect.

Science strives to resolve disputes dispassionately. It brings objective evidence to bear on fraught questions. In the absence of evidence, disputes get resolved by social status: powerful actors wield influence on public opinion and policy, while weaker ones get subjugated. Science aims to neutralize this personal power and to outsource the determination of truth to objective methods. It aims to do away with eminence-based influence and replace it with evidence-based decision-making.

Psychology's claims to pure objectivity can seem idealistic, if not naïve. However, its *aspiration* to be objective carries substantial moral weight. Groups whose relative power descends from tradition rather than equity have much to fear from the disruptive impact of empirical scrutiny.

Consider the example of socially conservative views on sexuality. By and large, the conservative approach to sexuality has been signifi-cantly undermined by empirical investigation. Research has highlighted the healthy nature of same-sex relationships, the positive outcomes for children of having same-sex parents, and the benefits for society of anti-discrimination policies and legislation (Meyer, 2013). Socially con-servative advocacy groups are thus especially keen to discuss limitations of the methods used and the broader issue of non-replicability in psy-chology (e.g., Iona Institute, 2012). To them the claim that psychology is in crisis is very welcome news. When psychology research consistently clashes with your preferred worldview, it is consoling to learn that the entire field may be unreliable.

Conservatives have long been accused of attacking the integrity of sci-ence in an effort to impede science-based policies that run contrary to their values or interests (Shulman, 2006). Rather than advance discrete arguments that support alternative explanations, the strategy is to high-light ambiguity in the mainstream consensus. In other words, the aim is to crank up the crisis talk.

In recent years, psychology has been accused of exhibiting a 'liberal' political bias. Critics argue that this political monoculture has a distort-ing effect on the field and that more should be done to attract socially conservative people into psychology (Duarte et al., 2015). According to this critique, broader viewpoint diversity would improve psychology research. The implication is that conservatives are right to be concerned about psychology's approach to social issues, because so few psycholo-gists are themselves socially conservative.

It is certainly discernible that academics tend to be less conservative than the general population. But whether this represents a destructive bias is less clear. For one thing, socially conservative worldviews regular-ly verge on anti-science. Whether research quality would really benefit from more of this type of thinking seems quite questionable.

In addition, psychology's lean to the 'left' is usually diagnosed by Americans, and thus from a US perspective. It is worth bearing in mind that sociopolitical ideas vary from country to country. What Americans refer to as 'liberal' is relatively conservative in global terms (Hilbig and Moshagen, 2015). Moreover, the accusation of 'liberal bias' is less a philosophical observation and more an incendiary soundbite emanating from domestic US politics. Psychologists are just one of many professional groups to be dismissed as 'snowflakes' by American conservatives. Given its US origins, the charge must be seen as at least partly politically partisan. Given the localized nature of this political dynamic, it is unlikely that liberal bias constitutes a true crisis for global psychology.

Sometimes it is psychologists themselves who wind up the dial on psychology's crises. Not all psychologists are overjoyed at psychology's scientific bent. Reflecting its many overlaps with fields such as philosophy, sociology, hermeneutics, and cultural studies, psychology contains a range of methodological approaches. Some of these – such as critical theory or qualitative methods – foster active antipathy towards mainstream science. Adherents feel that human minds cannot be measured and that psychologists' subjectivity cannot be circumvented. The replication crisis, therefore, is grist to their mill: it allows these non-scientific psychologists to say, 'We told you so.'

But even psychologists who are dedicated to science can specialize in crisis talk. They feel duty-bound to catastrophize, because they want psychology to clean up its act. They see the field as dysfunctional and in need of radical remedy. To them, crisis talk is neither strategic nor cathartic. It is a form of direct action: the first step to dealing with a problem is to admit that you *have* a problem.

Where the tragic happens

It might not seem obvious, but in many respects professional academia is a lot like the music business. Talented practitioners spend professional time reducing hundreds of hours of toil to an end-product that take a few minutes to consume. They strive for their output to be purchased, downloaded, and (ideally) appreciated. But even if it is ignored, a true rock star must keep on keeping on. They work on several projects at once, aiming for prolific productivity over time. Collaboration is the key to success (now and again a featured rapper might be called upon to bolster the gravitas of an otherwise forgettable track). Some outputs will be unexpected triumphs, others out-and-out duds, but each will stand or fall on its own merits. Yesterday's hits can be a source of nostalgia, but today's street cred requires new material, critical acclaim, and on-going chart success.

The music business is highly competitive. Few performers achieve the success they envisaged in their dreams; most are left to ponder what might have been. Obscurity is not so much a risk as a reasonable level to aim for. Churning out enough middle-of-the-road product to keep food on the table is all that matters in the end. Everything is much less glamorous than it appeared from the outside. Nonetheless, the show must go on ...

Professional academics go through all these experiences, but with journal articles instead of pop songs. Their energy is expended on designing new studies, garnering resources to conduct them, analysing the resulting data, writing the relevant papers, and submitting the stuff to journals. Little of this work is itemized on a paid-by-the-hour basis. Rather, it is absorbed into daily life, whenever and wherever it can be fitted in. Ultimately what counts is the success of the end-product. Formally and informally, the fortunes of academics are determined by research outputs. By their papers shall you know them. Reputations, promotions, introductions, and opportunities all flow from scholarly productivity.

In university life the slippery pole can be climbed in many ways: serving time on committees; navigating, or cutting through, bureaucracy; learning to decode criteria; knowing the right people; or – let's not forget – being an effective teacher. However, not all promotions are the same. The truest cachet attaches to those who are elevated on the basis of research. Nobody became an academic star by being good at chairing meetings. When a keynote speaker is announced, their bio will not tell us how many hours per week they teach. What counts as currency in this crazy, mixed-up, rock-'n'-roll world is volume of peer-reviewed publication.

The entire academic journal industry owes its origin to status anxiety in academia. Before the seventeenth century scholars and scientists published their ideas in books. Books were difficult to manufacture and slow to produce. If you wished to report something, your audience would have to wait to read about it. The delay meant that one of your rivals might get there first. If another scientist made the same discovery – but had better access to printing – they could publish their work before yours was ever read. Nobody would care that you were really the first with the finding. History would not even notice you.

Journals were established to deal with this problem. By containing works of several scientists, curtailing the length of articles, and appearing frequently throughout the year, journals offered a rapid and accessible way to disseminate discoveries. More importantly, they helped scientists establish primacy: they recorded who was first to produce particular findings. So it was that journals were invented to ease the angst of academics who feared falling behind.

Today, academic journal publishing is an enormous commercial industry, largely operated by private corporations. The market for science journals alone is believed to be around €10 billion per year (Ware and Mabe, 2015). Currently, 28,000 English-language journals publish 50,000 articles per week. According to the APA's PsycINFO database, around 2,000 journals publish psychology research. By rough approximation, this means that around 500 psychology articles are published each day.

The vast bulk are written by status-anxious academics. Each year, hundreds of thousands of researchers, most of them public servants, expend millions of pro bono hours producing material for journals to publish. They do this because, in one form or another, they want to advance their careers. For many, it's a simple matter of clocking up a paper count: the higher the number, the higher their score in the system that determines their rank.

True, for other academics publication in journals is a means to a different end: a way of sharing knowledge, solving problems, or advocating for positive change. However, even these colleagues will want their names on the papers. It matters to them that the world knows it is they who have authored their work. They may not be hungry for promotion, but they still infer status from publication.

Journal publishing is a highly profitable business. Even though subscription fees are high, most of the production is done by volunteers. Authors submit papers to editors, editors allocate submissions to associate editors, associate editors recruit peer reviewers, peer reviewers review submissions and return comments, associate editors pass these comments back to authors, and authors then rewrite their papers. For nearly every journal these tasks are done for free; nobody gets paid.

In fact, for an increasing number, the *opposite* is the case. Should a paper be deemed good enough to be published, then instead of being offered a royalty the author will receive a *bill*. An 'article processing fee' is required to ensure that the paper sees the light of day. Around 10 per cent of journals operate such pay-to-play systems (Ware and Mabe, 2015), where the author must cough up a thousand or so dollars or else ditch the paper from their résumé.

Each year more and more new journals are launched, the majority on a for-profit basis. Given that most private corporations run their businesses on commercial terms, there inevitably arises a question about how – or whether – quality control is ensured. At face value it seems that a worldwide public education sector is subsidising a worldwide private publishing industry. Universities pay for bulk subscriptions using taxpayers' money, while public servants supply the content for free, sometimes contributing a stipend for it to be published. If anything, the incentives in this industry are for *volume*, not quality.

In psychology, as for most sciences, academic journals are integral to the production and dissemination of knowledge. New research is not considered credible until it appears in a journal. Journals are the gatekeepers of the field's accumulating content. They determine what is and is not psychology. The merit of the field is shaped by their policies and practices. Therefore, if psychology is in crisis, then we need to talk about journals.

When journals sneeze, psychology catches a cold

Quality control in academic journals is managed through the peer review system. Each submitted paper is assigned to an associate editor, who co-ordinates the process. Their first task is to find some peer reviewers. This usually involves searching online for academics who are knowledgeable in the requisite subject area. Often several must be contacted before any agree to serve. Once two or three are recruited, each will read the submitted paper and write a report for the journal. They can reject the paper immediately or recommend it for further consideration. If the latter, their reports will contain suggestions as to how the paper might be improved. The authors are sent the reports and invited to reconfigure their paper to take account of the reviewers' recommendations. The paper is then resubmitted and reviewed all over again. There is no guarantee of eventual publication: if the modified manuscript is unsatisfactory, the reviewers will return it with fresh reports. Most papers undergo multiple such rewrites. A large number are still rejected in the end.

Rejection rates vary from journal to journal. Prestigious journals might turn down 90 per cent of submissions. A journal considered reasonably high quality could reject 75 per cent. Journals that charge 'article processing fees' usually have lower rejection rates but will still dump around a third of what they receive.

In principle, the system should work: it is blind, expert, conscientious, and balanced. Getting published is certainly not easy. However, practice does not always emulate principle. In practice the peer review system is not really blind, not really expert, not really conscientious, and not really balanced. Although a large portion of submissions are rejected in the name of quality control, the accuracy of the cull is open to question.

There are several reasons why the system is not really blind. Reviewers regularly realize whose papers they are reviewing. They recognize the methods of research, the laboratories described, the writing styles, or the ideas expressed. A frequent giveaway lies in the manuscript's bibliography: authors often overemphasize their own research when explaining new studies, creating clusters of conspicuous self-citations. Knowing whose work you are reviewing strains your objectivity. You might be sympathetic towards authors you know and cagey towards ones you've never heard of. You might allow grudges or rivalries to cloud your judgement. You may succumb to a range of unconscious biases prompted by the university or country the authors hail from. For such reasons many journals dispense with blinding altogether: reviewers remain anonymous, but author names are included with manuscripts. Editors argue that, as blinding doesn't really work, it is better to standardize the way that authors' identities become known.

The peer review system also falls short on expertise. Reviewer expertise cannot be ensured because it is not closely assessed. Journals rely on a range of dodgy indicators of reviewer knowledge. One example is publication record: reviewers are assumed to know a subject if they have previously authored (or coauthored) papers relating to it. However, as we will see shortly, coauthorship can accrue from many reasons *other* than subject-area awareness. Some journals try to divine expertise from reviewer self-report. On the journal's website potential reviewers are invited to tick boxes indicating their chosen specialist subjects. This approach is reliant on said reviewers having sufficient personal insight to recognize the limits of their own competence, a quality not shared widely among the human race in general. Another idea has been to ask authors to nominate their own reviewers. Perhaps inevitably, unscrupulous researchers have been known to try to game such systems. Some nominate their friends. Others invent fake reviewers or provide fake email addresses when nominating authentic ones. Ultimately, the peer review system does not *necessarily* ensure that reviewers have appropriate expertise. Many submissions – possibly the majority – are reviewed by non-experts.

There is also the matter of reviewer diligence: the peer review system is not inevitably conscientious. Nothing guarantees that reviewers will concentrate when writing reviews. If anything, the highly automated, deadline-driven nature of most journals' reviewer management systems is likelier to provoke irritation, and thus carelessness. The torrent of automatic emails that reviewers receive from journals can seem relentless. First you are invited to be a reviewer, then reminded you have been invited, then sent a note to thank you for accepting, then reminded you have accepted, then prompted to submit your review on time, then reminded that the deadline is approaching, then told that the deadline has passed, then sent a second reminder, and so on... While a barrage of emails from 'no-reply@journal.com' probably prevents you from *forgetting* your assignment, you might be sorely tempted to rush the job just to stem the flow of robotic correspondence.

Finally, the peer review system is not that well balanced. Some papers receive an in-depth and forensic assessment; others receive a lighter touch. Some reviewers concentrate on the overall concept; others get hung up on the smallest details. Some reviewers present their review in one hundred words; others require a thousand to deal with the same manuscript. Some reviewers write in a friendly and collaborative tone, even when rejecting a paper; others are austere and sanctimonious, even when recommending publication. A lot of variation results from inconsistencies in reviewer expertise and conscientiousness, as we have just discussed. However, much of it stems from the fact that reviewers rarely receive clear or consistent instructions from journals. Usually there is a permissive atmosphere: reviewers are trusted to decide for themselves how to produce their reviews. While reviews tend to be written in a particular style, this reflects more an evolved academic dialect than an agreed industry standard. There is no 'official' way to write a review and no ideal or definitive format.

Psychologists will be well aware of another factor that impinges on peer review: namely, the social context of author-reviewer relationships. All interactions between authors and reviewers amount to social exchanges characterized by power differentials. When a reviewer asserts their expertise, the author must respond in a way that meets the reviewer's needs as well as their own. Given the difference in status between subjugated author and pre-eminent referee, this usually involves much deference and forelock-tugging. In their cover letters authors will express effusive gratitude for the blessings bestowed upon them (e.g., 'We thank the reviewers for their helpful comments, which have greatly improved our paper'). Most will calculate that reviewers have a soft spot for personal flattery ('We thank the reviewers for their valuable insights'). If the reviewer has recommended that a specific point be added, the authors will transcribe it verbatim into their revised manuscript and then draw attention to having done so ('We thank the reviewer for this most helpful suggestion'). Should they find the feedback unfair, authors will do everything they can to disguise it. It is much safer to play the role of grateful serf, happy to facilitate the whims of one's social superior.

This culture of obsequiousness has arisen because it is adaptive. Authors are more likely to succeed if they do these things. For their part, reviewers are just as likely to succumb to this norm of social harmony. When things go well, the reviewer's amenability is reinforced by the author's praise, and the author's praise is reinforced by the reviewer's amenability. As love-bombing campaigns go, this escalation between reviewers and authors effectively constitutes an arms race. In short, publication decisions are influenced by more than just scientific merit. Any belief that peer review filters out flaws and polishes perfection should be dispensed with.

Its warts-and-all reality can be revealed by a thought experiment.

Imagine a paper has been submitted to a journal. After it is reviewed, the authors modify the manuscript and resubmit it in revised form. Assuming the recommendations have been acted upon, the paper is ultimately accepted. However, for the purposes of our thought experiment, it is not actually published. Instead, *the final version of the accepted paper is now resubmitted*, as though it were a brand new paper, to the same journal. It is assigned to a different associate editor who recruits a new set of reviewers. We should bear in mind that the version of the paper they read has already been declared fit for immediate publication by the journal's peer review process. But we can be pretty sure of the result of our thought experiment: the second set of reviewers will pick apart the manuscript, identify shortcomings, and write serious reports that include recommendations for adjustment. In their minds, the paper will *not* be fit for immediate publication. Prior to interacting with the authors, the reviewers will invariably assume the manuscript to be deficient. The reviews they write will confirm these self-fulfilling prophesies.

Our thought experiment leads us to a stark conclusion: no paper ever published in any journal would be considered ready for publication were it to be reviewed, as fresh, by new reviewers. In that sense, *no published paper meets the publication standards of the publication in which it is published.*

The entirety of psychology's subject area content – for example, every substantive point made in a textbook – is judged by the standards of journal peer review. Psychology studies are not considered credible unless they have appeared in peer-reviewed journals. We might assume that most publishers, editors, associate editors, and reviewers are honest people with high standards who want to do a good job. The problem is not that there is a clear preponderance of untrustworthy people involved. The problem is that bad practices are *possible*. If malpractice is not really controlled, then it is not really curtailed. It is impossible to be sure when and where we can have confidence in published research. Inconsistent reviewer selection, unjudged reviewer expertise, variable reviewer effort, unpredictable reviewing styles, and the social psychology of author-reviewer interactions do not bode well for the rigour of peer review as a system of quality control.

And yet this is the system that controls the quality of psychology.

Predator versus prey

Quality control can operate in top-down or bottom-up ways. Peer review represents the top-down approach: a community of regulators mobilized to police psychology's research supply chain. However, given its inconsistencies, peer review might be the supply chain's *weakest* link. The problem is compounded by the fact that its weaknesses are not always conspicuous. Indeed they are virtually invisible. When reading a paper, it is difficult – if not impossible – to gauge how robust the review process has been.

This may be why recent years have seen the emergence of so-called predatory publishers: fly-by-night commercial operations that purport to offer peer-reviewed publication but that really don't. Instead they are in the business of extracting article processing fees from gullible researchers. Typically, predatory journals will tempt potential authors with promises of swift publication. Many such journals have impressive-sounding names and decent-looking websites. But when it comes to a review process all corners are cut: the most undemanding of reviews (or none at all) will be produced. Virtually everything submitted ends up accepted for publication.

Even if only a handful of authors take up the offer each month, that still produces a handful of thousands of dollars' revenue. And in reality such journals deal with far more than handfuls (Kolata, 2017). Around the world, there are more than enough desperate academics to make these businesses extremely lucrative. According to one assessment, there are well over 1,000 predatory journals, with many specializing in psychology (Voutier, 2017). Experienced researchers are unlikely to be taken in by such scams. But not every university promotions board or tenure committee – many of which are staffed by bureaucrats and bean counters – will be so knowledgeable.

Top-down quality control, as represented by peer review, is insufficient to get the job done. Therefore, psychology needs quality control of the bottom-up kind. We need psychologists to cultivate their own culture of quality-promoting practices. We need them to police themselves.

So how is that going?

I wrote this

The production of published articles is the bread and butter of academic life. It is how researchers build their careers. But, as we have considered throughout this book, psychologists are mere human beings. They betray classic human frailties. While most will undoubtedly follow their journey with impeccable honesty and righteousness, some will surely slip along the way. When psychology researchers are left to police themselves, the resulting crime rate suggests an unfortunate tolerance of sundry misdemeanours.

One recurring ethical concern is the way in which authorship of collaborative work is determined. In most sciences, psychology included, research is conducted by teams. This means that nearly every paper represents a joint effort, with multiple coauthors. When determining coauthorship, the ethical requirements set out by the International Council of Medical Journal Editors (ICMJE, 2016) are seen as definitive. According to these, coauthorship should involve substantive input into the idea for the work (e.g., the research question), the accumulation of the work (e.g., gathering and analysis of data), *and* the consideration of the work (e.g., interpretation of data). There should *also* be direct involvement in the production of the paper, such as a share of the writing load or eyeball-level approval of the final written product. And there should *also* be a willingness to be accountable for the work, to stand up for its quality and respond to questions that might be raised about it. If a person is not contributing in *all* these ways, they should not be listed as a coauthor.

However, it is well known that attribution of authorship does not always adhere to these principles. A widely encountered problem is the inclusion of coauthors for political or diplomatic reasons. Some researchers habitually list their senior professors, team leaders, or other high-status individuals as coauthors on every paper they work on, regardless of whether those persons have had the required level of involvement. It is of course possible to organize a team such that its director is genuinely involved in every paper its members produce. However, in some large research groups senior figures are routinely listed as coauthors of papers that they actually know nothing about. Sometimes this happens because junior researchers wish to flatter senior colleagues. But frequently it happens because senior figures have made it clear that they *expect* it to happen: they see it as their entitlement to be listed as coauthors on all papers in their midst, whatever their level of input (if any).

Occasionally senior colleagues will rationalize this practice. They will say that even a passing influence on a junior researcher's thinking is sufficient to warrant coauthorship of their papers. A common assumption is that simply being a supervisor (of, say, a PhD student) or a boss (of, say, a professional research scientist) will produce such an entitlement. However, authorship refers to the creative process that leads to intellectual output, not to the bureaucratic process that governs the registration of students or employees. Coauthorship requires substantive intellectual activity. It is based on collaboration, not coercion. Simply being a supervisor or a boss does not make you the author of somebody else's writing.

Neither does having a peripheral or marginal role in the research. Another common controversy is the inclusion of vast numbers of coauthors on otherwise finite research papers. In some research areas it appears now almost conventional to include a complete dramatis personae of personnel attached to a project, similar in scope to the production credits that roll at the end of a movie or TV show. Thirty or forty people can be said to have contributed intellectually to a three- or four-page paper. In neuroscience there will frequently be more coauthors on the report than there are participants in the research being reported.

A particular problem relates to mass collaboration, especially in multi-centre studies where data are gathered across a range of different sites. When a large study involves, say, twenty different universities, there will usually be (at least) one investigator in each who can claim involvement in the project. However, whether each can claim involvement with the development of the research idea or the interpretation of findings is much less clear. More likely, their role will have been limited to collecting data using a preordained method and emailing a spreadsheet to the lead researcher before a set deadline.

Of course, it is beneficial to psychology that large-scale samples are studied. Psychology's statistical power and sampling reach both need to be enhanced. Therefore, all other things being equal, multi-site studies are inherently advantageous over single-site ones. However, the quality of the finished product – the peer-reviewed journal article – is also vital. Having a large number of 'passenger' coauthors free-riding on the article's byline is just not healthy.

For one thing, garnering coauthorship on the back of minimal contribution yields disproportionate credit – and thus unfair advantage – in subsequent employment or promotion processes. It distorts any competition in which number of publications is a favouring factor (e.g., when applying for jobs or research grants). It stands to reason: getting something for nothing, or for very little, is *by definition* a profitable exchange. The question is, who is incurring the loss? Simply put, the loss is incurred by all those researchers who apply for jobs or grants but who are *unwilling* to extract unwarranted credit from other people's intellectual work. Of course, such honesty might in time prove self-destructive. Passenger authorship is pernicious: it serves to inflate the *perceived* norm for research productivity in a given field (Moffatt, 2011), thereby creating a communal pressure on even more researchers to play along.

More simply, the inclusion of undeserving coauthors dilutes the relative contributions of the real authors. A doctoral student who works on several drafts of a paper only to find, at the last minute, that their supervisor has added some senior buddies as coauthors, might rightfully worry about sharing the limelight. What should be recognizable as chiefly their own work will forever be seen as a communal effort. On the other hand, the practice also aids those who *want* to camouflage their contribution. Authors with industry ties or blatant conflicts of interest may be only too glad to blend in with the crowd on a bloated byline (Panter, 2017).

Sometimes famous researchers are invited to be coauthors because the actual author wishes to glean some reflected glory from linking their paper to a prestigious laboratory. As we have seen, peer reviewers are liable to be swayed by such information when evaluating the quality of submitted manuscripts. So not only does this practice distort the public record regarding the study's provenance, it also produces an unfair publication advantage for the de facto author (offset, as ever, by the relative disadvantage faced by those who play by the rules).

Finally, lax approaches to coauthorship emit a mixed message about psychology's attitude to precision, accuracy, and transparency. As psychological science is largely self-policing, its long-term reputation depends on public trust in its integrity. It simply does not look good if psychologists are insincere about who exactly is responsible for what is said on the pages of journals. It leads to many awkward questions. If psychologists are willing to fudge the line-up of coauthors on their papers in order to portray a preferred version of reality to the world, then what else might they be fudging? And if they are willing to disregard a conventional principle of ethics that is supposed to govern and protect their work, then what other ethical principles do they habitually ignore?

The problem of misaligned incentives

Academia doesn't provide accolades to just anybody. Career progression is not based on being well dressed or having the neatest office. Permanence, promotions, and professorships go to those with the best publication statistics. Beyond the mere number of papers published, various metrics help co-ordinate the rat race. Each is as reductive as it is seductive.

One simple measure is the total number of citations received; in other words, the number of times a person's publications have themselves been listed in the bibliographies of other papers. A related metric is the *i*10-index: the number of papers in a person's résumé to have attracted ten or more citations. But perhaps the archetypal yardstick is the *h*-index. This is the number of papers a person has published that each have been cited that same number of times or more (e.g., if they have published five papers that have attracted five or more citations, their *h*-index will be five; if they have twenty papers with twenty or more citations, their *h*-index will be twenty; etc.).

In short, there are many methods for measuring psychologists' academic usefulness. The higher your metrics, the more useful you are: the more likely you are to be employed, made permanent, or promoted.

You might notice that such indices do not measure the *quality* of work produced. Occasionally, a researcher's oeuvre will go viral because critics have spotted egregious errors or other controversial content. That person can expect to receive many citations precisely because their work is used so repeatedly as an example of something awful. Over time, they will experience a healthy engorgement of their indices. Indices don't just decide personal promotions, they also contribute to university world rankings. Therefore, as far as a university is concerned, bad research that attracts citations is more useful than good research that doesn't.

Even the way in which *quantity* is indexed can be ropey. It is an established convention that shared authorship is counted as equal to full authorship. If a team of researchers have coauthored a paper, it counts as 'one' full paper in each of their résumés. Each citation received counts as 'one' full citation for each coauthor. In international rankings a university is said to have attracted one full citation every time a paper that one of its faculty has coauthored is cited. Occasionally an otherwise obscure institution will rise in the rankings because one of its employees had the good fortune to join a research consortium that produced a landmark article.

For example, in physics the 2015 paper that estimated the mass of the Higgs Boson particle had over *five thousand* coauthors (Castelvecchi, 2015). Some of the smaller universities involved enjoyed a discernible rankings boost caused by the citation spurt that ensued from this historic publication. Perhaps a more rational way to process the data would have been to log each coauthor as producing one five-thousandth of a paper that year. Fractionalizing would facilitate the aggregation of research productivity across any group of researchers, yet it is rarely done. Apparently, fractions are considered too mathematically complicated for auditing the research output of particle physicists (or indeed any other scientist).

What is good for a researcher is not necessarily good for research (Nosek, Spies, and Motyl, 2012). In fact, the opposite is more likely true: what is good for researchers can be materially *bad* for research. Career-wise, cutting corners is lucrative. *Not* cutting corners – refining your hypotheses, measures, methods, statistics, sampling, and so on – takes time. With more and more researchers competing for fewer and fewer faculty positions, taking time means falling behind. Modern academia is not a place for life's tortoises.

Academic incentive structures reward inflated quantity over actual quality, social loafing over self-sacrifice, and expedient, rather than ethical, accountability practices. The aim of the game is to produce statistically significant results, whatever their truth value (Smaldino and McElreath, 2016). The system encourages free-riding and a tolerance of error: the most important thing is to get your name on as many papers as possible and not worry too much, if at all, about replicability. Papers are your main concern; replicability can be other people's.

Journal publication, conducted in the context of global academia, is the sole method of producing legitimate subject-area content in psychology. As such, the future of the field seems in very unsafe hands. The production line is faulty. Quality control is porous. The incentives are all wrong.

On the face of things, psychology's crises are not going anywhere any time soon.

Responding to psychology's crises

All this crisis talk might leave the reader somewhat despondent. But despondency is a good thing. It is adaptive. A sense that hope is diminishing at least means there is still hope. A belief that things cannot go on like this means that we are more likely to ensure that they don't. Feeling blasé would do little to steel the soul. Talking – nay, obsessing – about crisis is a symptom of collective vigilance, a sign that we are unhappy with the flimsy present and aspire to an improved future. Our objective should not be to shore up psychology's reputation, to help psychologists save face, or to protect psychology's position in society. Nor should we try to avoid upset. Quite the contrary. Psychology's current state of homoeostasis is insidiously self-corrupting. It is our obligation to upset it.

Decorum is a double-edged sword, often deployed deviously to protect the status quo. True, an avoidance of conflict ensures that at least some people are happier for longer. But some people, without knowing it, are happy about bad things. Addressing these things will make such people less happy. If scientists have moral beliefs, these surely include commitments to accuracy, transparency, and the avoidance of harm. Systematic cross-cutting crises that lead psychology to promulgate erroneous findings, to obfuscate when questions are raised, or to defend poorly grounded practices should therefore be viewed as moral challenges. The emotional upset caused to a few psychologists should be factored into a broader moral calculus.

Quite often warnings about psychology's crises are met with a counter-allegation: the critics are exhibiting poor manners. As described earlier, debunkers of dodgy research have found themselves denounced as 'methodological terrorists' and 'shameless little bullies'. They are depicted as callous or partisan (Petrie and Weinman, 2017), and accused of inciting witch-hunts and corralling lynch mobs (Spellman, 2015). The fact that such characterizations are themselves inflammatory highlights the subjective nature of civility. One person's fair punch is another person's low blow. To argue that criticizing research should itself be criticized on the grounds that public criticism is uncalled-for is a particularly blind form of self-contradiction. Branding your detractors 'terrorists' or 'bullies' or dismissing their complaints as vexatious is as much an effort at public shaming as anything you might be upset about. (The related suggestion that criticism should itself be peer reviewed is, to put it mildly, missing the point.)

Perhaps it is revealing that at least some psychologists expend greater effort policing the tone of debate than the calibre of research. But as offended as they might be by a forthright critique, it does not absolve them from responding. A true act of decorum would be to rise above personality clashes and address accusations directly, regardless of how they are raised. As well as demonstrating gravitas, it is also prudent. Cognitive psychologist Tal Yarkoni (2016) has noted that if somebody is rude when telling you your car is on fire, it is still wise to put out the fire. It is decidedly *unwise* to delay doing so in order to rebuke the tenor of the warning.

What to do?

It is perhaps self-evident that addressing psychology's crises will require wholesale change in what it is psychologists do. Matters would be greatly improved were psychologists to study larger samples (thus boosting statistical power), publish *all* their findings (including studies that have 'failed'), report *all* their analyses (including power analyses), focus on effect sizes (instead of statistical significance), use measurements known to have good validity and reliability, and conduct research on wider cross-sections of the human race. The field would be markedly enhanced if psychologists conducted more replications and exercised due scepticism when citing research that has not yet been replicated. It would be lovely if they avoided exaggerating the impact of psychological research and wonderful if they concentrated harder when reviewing papers for journals. Additional education in research methods would be no bad thing, as would training in critical reading and practical communication skills.

While we are at it, we might take this opportunity to recommend that psychologists smile more at people in the street, give money to charity, reduce their carbon footprints, check in on elderly neighbours, and campaign for world peace. In other words, we would all like things to be better in every way. It is no great achievement to say so.

The real achievement would be to get even one of these things to happen. Take for example the power problem. That psychology research is underpowered has been prominently pointed out for decades. However, exhortations that power be increased have fallen on the deaf ears of researchers, reviewers, editors, grant-awarding bodies, policymakers, textbook writers, educators, and, ultimately, readers. Most psychologists agree that something should be done. Few are inclined to volunteer to do it. It is an apparent case of bystander apathy: the more they see their peers animated by the problem, the *less* likely they are to offer assistance. In their minds it is inevitable the problem will be solved eventually – by someone else.

The core issue is that of misaligned incentives. Psychologists are like bees in a swarm. It *looks* as if they move about in unison, as if acting on a single massed intelligence. However, in a swarm each bee is motivated individually to meet its own survival needs. Psychologists are much the same. To change the behaviour of the profession as a whole, we require that the incentives facing each and every psychologist be recalibrated. We need to ensure that what is good for the individual psychologist really is good for psychological science too.

By their nature, organizations like universities are hard to modify. Individuals and institutions are intertwined in global networks of norms. The career ladder to professorship is deeply embedded, having taken centuries to construct. No institution will be keen to change this aspect of their ecosystem. The fear is that sacrificing long-established practices could send them spiralling away from the mainstream to no good end. The risk of early-adopter syndrome seems more off-putting than any potential first-mover advantage. Asking universities to change the basis on which they recruit and promote academic staff is a non-starter.

More achievable would be to alter the way in which journals do their business. After all, while most are published by private commercial companies, they are edited by everyday academics. Journal practices are shaped by academic attitudes. Journals *can* insist on good things. For example, many journals already recommend that authors 'pre-register' their studies. This means that researchers must identify in advance that they wish to conduct a study, what their hypothesis is, and how they propose to pursue the investigation. If applied rigorously, pre-registration makes HARKing (hypothesizing after the results are known) a thing of the past and quashes the chaos caused by Rampant Methodological Flexibility.

Journals can also require authors to supply full methodological and analytical details before papers can be published. Such material can be posted on the journal's website. There is no academic or scientific reason to limit the sharing of technical information. Page-based space restrictions are nothing more than a legacy of the hardcopy era, when journals were printed on paper that cost actual money.

In similar ways, journals have the ability to reward data sharing and pooling. When psychologists conduct research, it is to the advantage of science as a whole if they make their data publicly available. In this day and age researchers can easily be required to post full datasets online.

When they do so, other researchers can independently scrutinize their analyses and confirm or enhance (or debunk) their findings. Furthermore, when multiple similar spreadsheets are available to download, researchers can pool them to create supersized datasets with superior statistical power. In some respects this is an ideal way of pursuing behavioural science: large datasets, combining samples from different locations, analysed by multiple investigators.

Journals can also restructure the peer review system to address some of its shortcomings. Consistency in the style and focus of reviews could be enhanced by issuing systematic instructions to reviewers, perhaps based on checklists of what needs to be covered (Nosek, Spies, and Motyl, 2012).

Improving reviewer expertise sounds more difficult, but it would be a good idea to at least admit that this is a problem. One approach might be to remove the anonymity of reviewers. Publishing their names alongside articles would fundamentally alter reviewers' behaviour. Perhaps it might encourage strategic sycophancy or have a chilling effect on the review process. On the other hand, it might stimulate reviewers to give maximum attention to what they write in reviews (e.g., they may avoid straying beyond their expertise). It would also assist the scientific community to gauge the strength of the journal's review process. Anonymous review is currently conventional, but its days may be numbered. In our era of transparency, accountability, and vigilance against discrimination, it seems anachronistic to allow anonymous reviewers to influence the salary prospects of government-paid public servants.

The psychological factors that distort peer review can also be addressed. For example, some journals are rolling out 'methods-only' reviews, where reviewers are asked to accept or reject papers on the basis of study design alone. In these systems reviewers are not shown the results of the study until after they have lodged their recommendation. This aims to ensure that reviewers avoid favouring significant results over null findings and thus to reduce the file drawer problem.

Biases stemming from author-reviewer interactions can be obviated by rotating reviewer assignments between revisions. A revised paper need not be reviewed by the same reviewers who evaluated its first draft (fresh reviewers are unlikely to be swayed by flattery directed at their predecessors). Finally, it may be helpful for journals to post all reviewer reports on their websites. This way the wider public can assess for themselves the reasonableness of reviewers' recommendations.

In summary: in a reimagined journal system authors would pre-register their intention to conduct a specific study, at which point their proposed methods would be formally peer-reviewed. Later, they would submit a partial report of the study along with the study's dataset. The journal would recruit new peer reviewers to recommend final publication on the basis of the methods described, prior to knowing the results. The journal could ask separate reviewers to analyse the dataset according to the pre-registered protocol, to confirm the authors' own results. The final manuscript would be reviewed in the traditional fashion. Assuming revisions are requested and carried out, the second draft would be reviewed by fresh reviewers. Ultimately the paper would be published along with a note listing all the various reviewers who have been involved. In the digital version readers would be able to click through various hyperlinks to obtain the full dataset, the full methodology (including questionnaires and supplementary materials), and copies of all the reviewers' reports. Papers would be published digitally on a rolling basis, with hard copies available to anyone who is willing to pay for them.

The need to push

A frequent concern about such changes is that it is already hard enough to recruit psychologists to serve as peer reviewers. Adding to the volume or complexity of work required would make it even harder. This is a pragmatic objection rather than a principled one. Currently there are few incentives to serve as a reviewer, and yet people do it. If incentives were put in place, presumably more would.

One such incentive might be to facilitate reviewers' getting their own papers published. Psychologists who are active peer reviewers could be given preferential treatment when it comes to the speed with which their own papers are processed by journals. On average, it currently takes around one hundred days after submission for a paper to be accepted for publication (Powell, 2016). Many journals take substantially more time. This suggests that there is scope for expediting the processing of submissions, at least in some cases. Prioritizing the submissions of researchers who appear on a cross-industry database of active reviewers might provide a due reward to those who give their time and effort in this manner.

Many traditions of academic journals date back to the nineteenth century. By and large, proposals for improvement reflect the possibilities of the internet age. They capitalize on the limitlessness of online repositories, the fluidity of electronic databases, the ubiquity of personal online accounts and profiles, and the ease of instantaneous person-to-person communication. This intrinsically technological aspect may imply that overhauling the academic journal system will require, or reflect, a generational shift (Spellman, 2015). Psychologists beginning their careers today will be impatient to introduce such innovations; their longer-serving colleagues might find them somewhat terrifying.

The near limitless flexibility of the internet offers an even more radical way to disrupt the misaligned incentives of academic research production. If the current system is dysfunctional in wrongly rewarding researchers for producing quantity over quality and if anonymous peer review is an unsound basis on which to determine what is publishable, then one solution would be to allow everyone to publish everything they like *without* the barrier of review (Nosek and Bar-Anan, 2012). After all, the current system in part reflects the need to cull submissions because of limitations on space, a now obsolete concern. And peer reviewing predates the contemporary world in which people are able to post online comments immediately after work is published. If scientific research was posted to websites on the expectation that peers would be invited to post their critiques alongside them, then we might enter a new world of *post-publication* reviewing.

The removal of pre-publication reviews would kill the idea of publication as an end in itself. It would nullify the false accounting of researcher productivity. Instead, researcher performance would need to be evaluated differently. Incentives for individual researchers would shift from bean counting to providing evidence of research quality. Psychologists whose work is consistently applauded in post-publication review would be given greater rewards than those whose work is perennially panned.

The prospect of a virtual public square in which research psychologists self-police quality by normalizing scientific critique might seem rather naïve. To be sure, in other contexts, crowdsourcing does not always turn out well. Ultimately, we remain reliant on human integrity and honesty. A bold new world will be of little consequence without a commitment to the pursuit of truth. New systems won't amount to much unless there is a determination to make them work.

Psychologists need to recognize that their status quo is unsustainable. They need to acknowledge that their discipline faces many problems. And they need to turn their evaluative gaze inwards, to accept that critical thinking is nothing without the ability to self-criticize.

It is of course difficult to incentivize people to criticize themselves. Can we get people to admit that while they are wonderful and exciting and full of potential, they are also sometimes just half-cocked, self-centred, and parsimonious with the truth? Contrary to cliché, it is not misanthropic to emphasize flaws when depicting the character of human beings. True misanthropy is to assume that humans are robotically benign and will gravitate toward uniform purity if given enough time. If psychology teaches us anything, it is that people are a rich and varied bunch: responsive to environment and susceptible to error; aesthetically nuanced and ethically variegated; highly intelligent and endlessly capable. And as is repeatedly worth pointing out: contrary to delusion, psychologists are people too.

Psychology has a rich history as the scientific study of human welfare, work, and whimsy. We owe it to the world to accept our many crises – and to take serious action to address them.

References

Achenbach, J. (2015, August 27). *Many scientific studies can't be replicated. That's a problem*. Retrieved from Washington Post: www.washingtonpost.com/news/speaking-of-science/wp/2015/08/27/trouble-in-science-massive-effort-to-reproduce-100-experimental-results-succeeds-only-36-times/.

Ackerman, S. (2015, July 14). *Three senior officials lose their jobs at APA after US torture scandal*. Retrieved April 1, 2017, from Guardian: www.theguardian.com/us-news/2015/jul/14/apa-senior-officials-torture-report-cia.

American Psychiatric Association (2013). *Diagnostic and Statistical Manual of Mental Disorders* (5th ed.). Washington, DC: American Psychiatric Association.

Anderson, C. J., Bahník, S., Barnett-Cowan, M., Bosco, F. A., Chandler, J., Chartier, C. R., ... Zuni, K. (2016). Response to comment on 'Estimating the reproducibility of psychological science'. *Science, 351*(6277), 1037-c.

APA (2017, April 15). *Division profiles by division*. Retrieved from American Psychological Association: www.apa.org/about/division/officers/services/profiles.aspx.

Arnett, J. J. (2008). The neglected 95%: Why American psychology needs to become less American. *American Psychologist, 63*(7), 602–614.

Baker, M. (2015, August 27). *Over half of psychology studies fail reproducibility test. Nature*. Retrieved from Nature: www.nature.com/news/over-half-of-psychology-studies-fail-reproducibility-test-1.18248.

Bargh, J. A., Chen, M., and Burrows, L. (1996). Automaticity of social behavior: Direct effects of trait construct and stereotype activation on action. *Journal of Personality and Social Psychology, 71*(2), 230–244.

Bartlett, T. (2014, June 23). *Replication crisis in psychology research turns ugly and odd*. Retrieved April 1, 2017, from Chronicle of Higher Education: www.chronicle.com/article/Replication-Crisis-in/147301.

Bastian, H. (2016, December 5). *Reproducibility crisis timeline: Milestones in tackling research reliability*. Retrieved April 1, 2017, from PLOS Blogs: Absolutely Maybe: http://blogs.plos.org/absolutely-maybe/2016/12/05/reproducibility-crisis-timeline-milestones-in-tackling-research-reliability/.

Bateson, M., Nettle, D., and Roberts, G. (2006). Cues of being watched enhance cooperation in a real-world setting. *Biology Letters, 2*(3), 412–414.

Baumeister, R. F., Bratslavsky, E., Muraven, M., and Tice, D. M. (1998). Ego depletion: Is the active self a limited resource? *Journal of Personality and Social Psychology, 74*(5), 1252–1265.

Beall, A. T., and Tracy, J. L. (2013). Women are more likely to wear red or pink at peak fertility. *Psychological Science, 24*(9), 1837–1841.

Bech, P., Rasmussen, N.-A., Olsen, L. R., Noerholm, V., and Abildgaard, W. (2001). The sensitivity and specificity of the Major Depression Inventory, using the Present State Examination as the index of diagnostic validity. *Journal of Affective Disorders, 66*, 159–164.

Beck, A. T., Ward, C. H., Mendelson, M., Mock, J., and Erbaugh, J. (1961). An inventory for measuring depression. *Archives of General Psychiatry, 4,* 561–571.

Begley, C. G., and Ellis, L. M. (2012). Drug development: Raise standards for preclinical cancer research. *Nature, 483*, 531–533.

Bennett, C. M., and Miller, M. B. (2010). How reliable are the results from functional magnetic resonance imaging? *Annals of the New York Academy of Sciences, 1191*, 133–155.

Benson, K. E. (2007). Enhanced female brood patch size stimulates male courtship in *Xiphophorus helleri. Copeia, 2007*(1), 212–217.

Beyerstein, B. L. (1999). Whence cometh the myth that we only use ten percent of our brains? In S. Della Sala (Ed.), *Mind Myths: Exploring Everyday Mysteries of the Mind and Brain* (pp. 1–24). Chichester: Wiley.

Bishop, D. (2014, September 26). *Why most scientists don't take Susan Greenfield seriously.* Retrieved April 1, 2017, from BishopBlog: http://deevybee. blogspot.ie/2014/09/why-most-scientists-dont-take-susan.html.

Blake, A. (2017, January 22). *Kellyanne Conway says Donald Trump's team has 'alternative facts.' Which pretty much says it all.* Retrieved April 1, 2017, from Washington Post: www.washingtonpost.com/news/the-fix/wp/2017/01/22/ kellyanne-conway-says-donald-trumps-team-has-alternate-facts-which-pretty-much-says-it-all/.

Boekel, W., Wagenmakers, E. J., Belay, L., Verhagen, J., Brown, S., and Forstmann, B. U. (2015). A purely confirmatory replication study of structural brain-behavior correlations. *Cortex, 66*, 115–133.

Bowerman, M., and Choi, S. (2001). Shaping meanings for language: Universal and language-specific in the acquisition of spatial semantic categories. In M. Bowerman, and S. C. Levinson (Eds), *Language Acquisition and Conceptual Development* (pp. 475–511). Cambridge: Cambridge University Press.

Branch, M. (2014). Malignant side effects of null-hypothesis significance testing. *Theory and Psychology, 24*(2), 256–277.

Brauch, K., Pfefferle, D., Hodges, K., Möhle, U., Fischer, J., and Heistermann, M. (2007). Female sexual behavior and sexual swelling size as potential cues for males to discern the female fertile phase in free-ranging Barbary macaques (*Macaca sylvanus*) of Gibraltar. *Hormones and Behavior, 52*(3), 375–383.

British Psychological Society (2013). *'Accreditation through Partnership' Handbook: Guidance for Undergraduate and Conversion Programmes in Psychology,* September 2013. Leicester: British Psychological Society.

Burt, C. (1962). The concept of consciousness. *British Journal of Psychology, 53,* 229–242.

Button, K. S., Ioannidis, J. P., Mokrysz, C., Nosek, B. A., Flint, J., Robinson, E. S., and Munafò, M. R. (2013). Power failure: Why small sample size undermines the reliability of neuroscience. *Nature Reviews Neuroscience, 14,* 365–376.

Candia, V., Weinbruch, C., Elbert, T., Rockstroh, B., and Ray, W. J. (2003). Effective behavioral treatment of focal hand dystonia in musicians alters somatosensory cortical organization. *Proceedings of the National Academy of Sciences, 100,* 7942–7946.

Čapek, K., and Selver, P. (1923). *R.U.R. (Rossum's Universal Robots): A Fantastic Melodrama.* Garden City, NY: Doubleday, Page.

Carbon, C.-C., and Hesslinger, V. M. (2011). Bateson et al.'s (2006) cues-of-being-watched paradigm revisited. *Swiss Journal of Psychology/Schweizerische Zeitschrift für Psychologie/Revue Suisse de Psychologie, 70*(4), 203–210.

Carey, B. (2015, August 27). *Many psychology findings not as strong as claimed, study says.* Retrieved from New York Times: www.nytimes.com/2015/08/28/science/many-social-science-findings-not-as-strong-as-claimed-study-says.html.

Carney, D. R. (2016, September). *My position on 'power poses'.* Retrieved April 1, 2017, from http://faculty.haas.berkeley.edu/dana_carney/pdf_My%20position%20on%20power%20poses.pdf.

Carney, D. R., Cuddy, A. J., and Yap, A. J. (2010). Power posing: Brief nonverbal displays affect neuroendocrine levels and risk tolerance. *Psychological Science, 21*(10), 1363–1368.

Carter, E. C., Kofler, L. M., Forster, D. E., and McCullough, M. E. (2015). A series of meta-analytic tests of the depletion effect: Self-control does not seem to rely on a limited resource. *Journal of Experimental Psychology: General, 144,* 796–815.

Castelvecchi, D. (2015, May 15). *Physics paper sets record with more than 5,000 authors.* Retrieved April 1, 2016, from Nature News: www.nature.com/news/physics-paper-sets-record-with-more-than-5-000-authors-1.17567.

Chen, S. H., Weng, L. C., Su, Y. J., Wu, H. M., and Yang, P. F. (2003). Development of Chinese Internet Addiction Scale and its psychometric study. *Chinese Journal of Psychology, 45,* 279–294.

Chiao, J. Y. (2009). Cultural neuroscience: A once and future discipline. *Progress in Brain Research, 178,* 287–304.

Christensen, C. M., Baumann, H., Ruggles, R., and Sadtler, T. M. (2006). Disruptive innovation for social change. *Harvard Business Review, 84*(12), 1–8.

Chung, R. (2012). *The Freedom to Publish Opinion Poll Results: A Worldwide Update of 2012.* Hong Kong: University of Hong Kong.

Cohen, J. (1994). The earth is round (p < .05). *American Psychologist, 49*(12), 997–1003.

Colgan, S. L., Faasse, K., Pereira, J. A., Grey, A., and Petrie, K. J. (2016). Changing perceptions and efficacy of generic medicines: An intervention study. *Health Psychology, 35*(11), 1246–1253.

Connellan, J., Baron-Cohen, S., Wheelwright, S., Batki, A., and Ahluwalia, J. (2000). Sex differences in human neonatal social perception. *Infant Behavior and Development, 23,* 113–118.

Connor, S. (2015, August 27). *Study reveals that a lot of psychology research really is just 'psycho-babble'.* Retrieved from Independent: www.independent.co.uk/news/science/study-reveals-that-a-lot-of-psychology-research-really-is-just-psycho-babble-10474646.html.

Coon, D. J. (1992). Testing the limits of sense and science: American experimental psychologists combat spiritualism, 1880–1920. *American Psychologist, 47*(2), 143–152.

Cowling, D. (2016, January 20). *Should polling be banned before an election?* Retrieved April 1, 2017, from BBC News: www.bbc.com/news/uk-politics-35350419.

Cuddy, A. (2012, October). *Subtitles and transcript: Your body language shapes who you are.* Retrieved from TED.com: www.ted.com/talks/amy_cuddy_your_body_language_shapes_who_you_are/transcript.

Cupp, S. E. (2014, June 4). *Girl hurricanes get no respect.* Retrieved April 1, 2017, from CNN: http://edition.cnn.com/2014/06/04/opinion/cupp-hurricanes-with-female-names/.

Deer, B. (2011). How the case against the MMR vaccine was fixed. *BMJ, 342,* c5347.

Diener, E., Emmons, R. A., Larson, R. J., and Griffin, S. (1985). The satisfaction with life scale: A measure of life satisfaction. *Journal of Personality Assessment, 49,* 1–5.

Doyen, S., Klein, O., Pichon, C. L., and Cleeremans, A. (2012). Behavioral priming: It's all in the mind, but whose mind? *PLoS ONE, 7*(1), e29081.

Dragioti, E., Karathanos, V., Gerdle, B., and Evangelou, E. (2017). Does psychotherapy work? An umbrella review of meta-analyses of randomized controlled trials. *Acta Psychiatrica Scandinavica, 136*(3), 236–246.

Duarte, J. L., Crawford, J. T., Stern, C., Haidt, J., Jussim, L., and Tetlock, P. E. (2015). Political diversity will improve social psychological science. *Behavioral and Brain Sciences, 38*, e130.

Earp, B. D., Everett, J. A., Madva, E. N., and Hamlin, J. K. (2014). Out, damned spot: Can the 'Macbeth Effect' be replicated? *Basic and Applied Social Psychology, 36*, 91–98.

Edwards, J. (2017). PACE team response shows a disregard for the principles of science. *Journal of Health Psychology, 22*(9), 1155–1158.

Eklund, A., Nichols, T. E., and Knutsson, H. (2016). Cluster failure: Why fMRI inferences for spatial extent have inflated false-positive rates. *PNAS, 113*(28), 7900–7905.

Engber, D. (2017, September 5). *The Bush torture scandal isn't over.* Retrieved October 1, 2017, from Slate: www.slate.com/articles/health_and_science/science/2017/09/should_psychologists_take_the_blame_for_greenlighting_bush_era_enhanced.html.

Estes, S. (2012, December 20). *The myth of self-correcting science.* Retrieved from Atlantic:http:/www.theatlantic.com/health/print/2012/12/the-myth-of-self-correcting-science/266228/.

Etchells, P. (2015, January 16). *Declinism: Is the world actually getting worse?* Retrieved April 1, 2017, from Guardian: www.theguardian.com/science/headquarters/2015/jan/16/declinism-is-the-world-actually-getting-worse.

Fanelli, D. (2009). How many scientists fabricate and falsify research? A systematic review and meta-analysis of survey data. *PLoS ONE, 4*(5), e5738.

Faux, R. (2014). Quantitative research. In T. Teo (Ed.), *Encyclopedia of Critical Psychology* (pp. 1614–1618). New York: Springer.

Felmingham, K., Kemp, A., Williams, L., Das, P., Hughes, G., Peduto, A., and Bryant, R. (2007). Changes in anterior cingulate and amygdala after cognitive behavior therapy of posttraumatic stress disorder. *Psychological Science, 18*, 127–129.

Feynman, R. (1998). *The Meaning of It All: Thoughts of a Citizen-Scientists.* London: Penguin.

Fuchs, A. H. (2000). Contributions of American mental philosophers to psychology in the United States. *History of Psychology, 3*(1), 3–19.

Fulford, K. W. (1989). *Moral Theory and Medical Practice.* Cambridge: Cambridge University Press.

Gallup (2017, May 22). *In US, belief in creationist view of humans at new low.* Retrieved June 1, 2017, from Gallup News: http://news.gallup.com/poll/210956/belief-creationist-view-humans-new-low.aspx.

Gani, A. (2015, September 2). *Girl, 12, achieves maximum Mensa IQ test score.* Retrieved April 1, 2017, from Guardian: www.theguardian.com/society/2015/sep/02/girl-12-wins-maximum-mensa-iq-test-score.

Garrison, K. E., Tang, D., and Schmeichel, B. J. (2016). Embodying power: A preregistered replication and extension of the power pose effect. *Social Psychological and Personality Science, 7*(7), 623–630.

Gates, S., Nygård, H. M., Strand, H., and Urdal, H. (2016). *Trends in Armed Conflict, 1946–2014.* Oslo: PRIO.

Gelman, A. (2016, September 21). *What has happened down here is the winds have changed.* Retrieved April 1, 2017, from Statistical Modeling, Causal Inference, and Social Science: http://andrewgelman.com/2016/09/21/what-has-happened-down-here-is-the-winds-have-changed/.

Geraghty, K. J. (2017). 'PACE-Gate': When clinical trial evidence meets open data access. *Journal of Health Psychology, 22*(9), 1106–1112.

Gibson, E. J., and Walk, R. D. (1960). The 'visual cliff'. *Scientific American, 202,* 67–71.

Gilbert, D. T., King, G., Pettigrew, S., and Wilson, T. D. (2016). Comment on 'Estimating the reproducibility of psychological science'. *Science, 351*(6277), 1037-a.

Gilman, S. L. (2014). 'Stand up straight': Notes toward a history of posture. *Journal of Medical Humanities, 35*(1), 57–83.

Given, L. M. (2008). *The Sage Encyclopedia of Qualitative Research Methods.* Los Angeles, CA: Sage.

Golder, A. (2017, October 7). *17 facts that sound fake but are actually totally true.* Retrieved November 1, 2017, from BuzzFeed: www.buzzfeed.com/andyneuenschwander/17-facts-that-sound-completely-wrong-but-are-100-t.

Goudsmit, E., and Howes, S. (2017). Bias, misleading information and lack of respect for alternative views have distorted perceptions of myalgic encephalomyelitis/chronic fatigue syndrome and its treatment. *Journal of Health Psychology, 22*(9), 1159–1167.

Greenfield, S. (2014). *Mind Change: How Digital Technologies Are Leaving Their Mark on Our Brains.* London: Random House.

Gutchess, A. H., Welsh, R. C., Boduroglu, A., and Park, D. C. (2006). Cultural differences in neural function associated with object processing. *Cognitive, Affective and Behavioral Neuroscience, 6,* 102–109.

Hagger, M. S., Chatzisarantis, N. L. D., Alberts, H., Anggono, C. O., Batailler, C., Birt, A. R., ... Zwienenberg, M. (2016). A multilab preregistered replication of the ego-depletion effect. *Perspectives on Psychological Science, 11*(4), 546–573.

Haller, S., and Krauss, S. (2002). Misinterpretations of significance: A problem students share with their teachers? *Methods of Psychological Research, 7*, 1–20.

Hamilton, M. (1960). A rating scale for depression. *Journal of Neurology, Neurosurgery and Psychiatry, 23*, 56–62.

Heidegger, M. (1927). *Sein und Zeit.* Tübingen: Niemeyer.

Henrich, J. (2008). A cultural species. In M. Brown (Ed.), *Explaining Culture Scientifically* (pp. 184–210). Seattle, WA: University of Washington Press.

Henrich, J., Heine, S. J., and Norenzayan, A. (2010). The weirdest people in the world? *Behavioral and Brain Sciences, 33*, 61–83.

Hilbig, B. E., and Moshagen, M. (2015). A predominance of self-identified Democrats is no evidence of a leftward bias. *Behavioral and Brain Sciences, 38*, e146.

Hintsa, T., Jokela, M., Elovainio, M., Määttänen, I., Swan, H., Hintsanen, H., ... Keltikangas-Järvinen, L. (2016). Stressful life events and depressive symptoms among symptomatic long QT syndrome patients. *Journal of Health Psychology, 21*(4), 505–512.

Hoffmann, T. C., and Del Mar, C. (2015). Patients' expectations of the benefits and harms of treatments, screening, and tests. *JAMA Internal Medicine, 175*(2), 274–286.

Holmes, T. H., and Rahe, R. H. (1967). The social readjustment rating scale. *Journal of Psychosomatic Research, 11*, 213–218.

Huff, D. (1954). *How to Lie with Statistics.* New York, NY: Norton.

Hughes, B. M. (2008). How should clinical psychologists approach complementary and alternative medicine? *Clinical Psychology Review, 28*, 657–675.

Hughes, B. M. (2016). *Rethinking Psychology: Good Science, Bad Science, Pseudoscience.* London: Palgrave.

Hughes, B. M., and Lü, W. (2017). Blood pressure reactivity or responses. In M. D. Gellman, and J. R. Turner (Eds.), *Encyclopedia of Behavioral Medicine.* New York, NY: Springer.

Hume, D. (1798). *A Treatise of Human Nature: Being an Attempt to Introduce the Experimental Method of Reasoning into Moral Subjects,* Vol. I. London: John Noon.

Hunt, E. (2011). *Human Intelligence.* Cambridge: Cambridge University Press.

Hyde, J. S., Lindberg, S. M., Linn, M. C., Ellis, A. B., and Williams, C. C. (2008). Gender similarities characterize math performance. *Science, 321*, 494–495.

ICMJE (2016, December 1). *Recommendations for the conduct, reporting, editing, and publication of scholarly work in medical journals.* Retrieved April 1, 2017, from International Committee of Medical Journal Editors: www.icmje. org/recommendations/.

IISS (2017). *Armed Conflict Survey 2017.* London: International Institute for Strategic Studies.

Institute of Medicine. (2015). *Beyond Myalgic Encephalomyelitis/Chronic Fatigue Syndrome: Redefining an Illness.* Washington, DC: National Academies Press.

Ioannidis, J. P. (2005). Why most published research findings are false. *PLoS Medicine, 2*(8), e124.

Iona Institute (2012, August 31). *Gay marriage researcher cleared of scientific misconduct charge.* Retrieved April 1, 2017, from Iona Institute: www.ionainstitute. ie/gay-marriage-researcher-cleared-of-scientific-misconduct-charge/.

Izard, C. E., Dougherty, F. E., Bloxom, B. M., and Kotsch, N. E. (1974). *The Differential Emotions Scale: A Method of Measuring the Subjective Experience of Discrete Emotions.* Nashville, TN: Vanderbilt University Press.

James, J. E. (2016). *The Health of Populations: Beyond Medicine.* London: Academic Press.

Jason, L. A., Evans, M., Porter, N., Brown, M., Brown, A., Hunnell, J., … Friedberg, F. (2010). The development of a revised Canadian Myalgic Encephalomyelitis Chronic Fatigue Syndrome case definition. *American Journal of Biochemistry and Biotechnology, 6*(2), 120–135.

Jefferson, A., Bortolotti, L., and Kuzmanovic, B. (2017). What is unrealistic optimism? *Consciousness and Cognition, 50*(1), 3–11.

John, L. K., Loewenstein, G., and Prelec, D. (2012). Measuring the prevalence of questionable research practices with incentives for truth telling. *Psychological Science, 23*(5), 524–532.

Johnson, D. J., Cheung, F., and Donnellan, M. B. (2014). Does cleanliness influence moral judgments? Direct replication of Schnall, Benton, and Harvey (2008). *Social Psychology, 45*(3), 209–215.

Johnson, R. B., Onwuegbuzie, A. J., and Turner, L. A. (2007). Toward a definition of mixed methods research. *Journal of Mixed Methods Research, 1*, 112–133.

Jones, J. T., Pelham, B. W., Carvallo, M., and Mirenberg, M. C. (2004). How do I love thee? Let me count the Js: Implicit egotism and interpersonal attraction. *Journal of Personality and Social Psychology, 87*, 665–683.

Jung, K., Shavitt, S., Viswanathan, M., and Hilbe, J. M. (2014). Female hurricanes are deadlier than male hurricances. *Proceedings of the National Academy of Sciences, 111*(24), 8782–8787.

Kagan, J. (2013). Equal time for psychological and biological contributions to human variation. *Review of General Psychology, 17*(4), 351–357.

Karasik, L. B., Adolph, K. E.-L., and Bornstein, M. H. (2010). WEIRD walking: Cross-cultural research on motor development. *Behavioral and Brain Sciences, 33*(2/3), 95–96.

Kellaway, L. (2017, May 7). *There is nothing cute about innumeracy.* Retrieved May 7, 2017, from Financial Times: www.ft.com/content/3174d5ce-30e 7-11e7-9555-23ef563ecf9a.

Kerr, N. L. (1998). HARKing: Hypothesizing after the results are known. *Personality and Social Psychology Review, 2*(3), 196–217.

Kindlon, T. (2017). Do graded activity therapies cause harm in chronic fatigue syndrome? *Journal of Health Psychology, 22*(9), 1146–1154.

Kolata, G. (2017, October 30). *Many academics are eager to publish in worthless journals.* Retrieved November 1, 2017, from New York Times: www.nytimes. com/2017/10/30/science/predatory-journals-academics.html.

Kostyleff, N. (1911). *La crise de la psychologie expérimentale.* Paris: Alcan.

Kuhn, T. S. (1962). *The Structure of Scientific Revolutions.* Chicago, IL: University of Chicago Press.

Kuss, D. J., Griffiths, M. D., Karila, L., and Billieux, J. (2014). Internet addiction: A systematic review of epidemiological research for the last decade. *Current Pharmaceutical Design, 20*(25), 4026–4052.

Larousserie, D. (2016, April 5). *La psychologie est-elle en crise?* Retrieved from Le Monde: www.lemonde.fr/sciences/article/2016/04/05/la-psychologie-est-elle-en-crise_4896152_1650684.html.

Lazarus, R. S., and Folkman, S. (1984). *Stress, Appraisal, and Coping.* New York, NY: Springer.

Leeb, R. T., and Rejskind, F. G. (2004). Here's looking at you, kid! A longitudinal study of perceived gender differences in mutual gaze behavior in young infants. *Sex Roles, 50*(1/2), 1–14.

Leggett, N. C., Thomas, N. A., Loetscher, T., and Nicholls, M. E. (2013). The life of p: 'Just significant' results are on the rise. *Quarterly Journal of Experimental Psychology, 66*(12), 2303–2309.

Letzter, R. (2016, September 22). *Scientists are furious after a famous psychologist accused her peers of 'methodological terrorism'.* Retrieved April 1, 2017, from Business Insider: http://uk.businessinsider.com/susan-fiske-methodological-terrorism-2016-9.

Lewis, J., and Lewis, B. (in press). The myth of declining violence: Liberal evolutionism and violent complexity. *International Journal of Cultural Studies.* DOI: 1367877916682108.

Liebig, J., Peeters, C., Oldham, N. J., Markstädter, C., and Hölldobler, B. (2000). Are variations in cuticular hydrocarbons of queens and workers a reliable signal of fertility in the ant *Harpegnathos saltator*? *PNAS, 97*(8), 4124–4131.

Lilienfeld, S. O. (2012). Public skepticism of psychology: Why many people perceive the study of human behavior as unscientific. *American Psychologist, 67*, 111–129.

Lin, F., Zhou, Y., Du, Y., Qin, L., Zhao, Z., Xu, J., and Lei, H. (2012). Abnormal white matter integrity in adolescents with internet addiction disorder: A tract-based spatial statistics study. *PLoS ONE, 7*(1), e30253.

List of psychological schools (2018, February 23). Retrieved February 23, 2018, from Wikipedia: https://en.wikipedia.org/wiki/List_of_psychological_schools.

Lopez, S. R., and Guarnaccia, P. J. (2000). Cultural psychopathology: Uncovering the social world of mental illness. *Annual Review of Psychology, 51*, 571–598.

Lopizzo, N., Tosato, S., Begni, V., Tomassi, S., Cattane, N., Barcella, M., ... Cattaneo, A. (2017). Transcriptomic analyses and leukocyte telomere length measurement in subjects exposed to severe recent stressful life events. *Translational Psychiatry, 7*, e1042.

Lubet, S. (2017). Investigator bias and the PACE trial. *Journal of Health Psychology, 22*(9), 1123–1127.

Luck, S. J., and Gaspelin, N. (2017). How to get statistically significant effects in any ERP experiment (and why you shouldn't). *Psychophysiology, 54*, 146–157.

Lynott, D., Corker, K. S., Wortman, J., Connell, L., Donnellan, M. B., Lucas, R. E., and O'Brien, K. (2014). Replication of 'Experiencing physical warmth promotes interpersonal warmth' by Williams and Bargh (2008). *Social Psychology, 45*(3), 216–222.

Madrigal, A. (2009, September 9). *Scanning dead salmon in fMRI machine highlights risk of red herrings*. Retrieved April 1, 2017, from Wired: www.wired.com/2009/09/fmrisalmon/.

Majid, A., and Levinson, S. C. (2010). WEIRD languages have misled us, too. *Behavioral and Brain Sciences, 33*, 103.

Makel, M. C., Plucker, J. A., and Hegarty, B. (2012). Replications in psychology research: How often do they really occur? *Perspectives on Psychological Science, 7*(6), 537–542.

Mance, H. (2016, June 3). *Britain has had enough of experts, says Gove*. Retrieved April 1, 2017, from Financial Times: www.ft.com/content/3be49734-29c b-11e6-83e4-abc22d5d108c.

Marshall, A. C., Cooper, N. R., and Geeraert, N. (2016). Experienced stress produces inhibitory deficits in old adults' Flanker task performance: First evidence for lifetime stress effects beyond memory. *Biological Psychology, 113*(1), 1–11.

Marshall-Gradisnik, S., Johnston, S., Chacko, A., Nguyen, T., Smith, P., and Staines, D. (2016). Single nucleotide polymorphisms and genotypes of transient receptor potential ion channel and acetylcholine receptor genes from isolated B lymphocytes in myalgic encephalomyelitis/chronic fatigue syndrome patients. *Journal of International Medical Research, 44*(6), 1381–1394.

Maryanski, A. (2010). WEIRD societies may be more compatible with human nature. *Behavioral and Brain Sciences, 33*(2/3), 103–104.

McCabe, D. P., and Castel, A. D. (2008). Seeing is believing: The effect of brain images on judgments of scientific reasoning. *Cognition, 107*, 343–352.

McCook, A. (2016, March 24). *Retractions rise to nearly 700 in fiscal year 2015 (and psst, this is our 3,000th post)*. Retrieved April 1, 2017, from Retraction Watch: http://retractionwatch.com/2016/03/24/retractions-rise-to-nearly-700-in-fiscal-year-2015-and-psst-this-is-our-3000th-post/.

McEvedy, C. P., and Beard, A. W. (1970). Concept of benign myalgic encephalomyelitis. *British Medical Journal, 1*(5687), 11–15.

McPhee, G. (2017). Cognitive behaviour therapy and objective assessments in chronic fatigue syndrome. *Journal of Health Psychology, 22*(9), 1181–1186.

Meehl, P. E. (1967). Theory-testing in psychology and physics: A methodological paradox. *Philosophy of Science, 34*, 103–115.

Meltzoff, A. N., and Moore, M. K. (1977). Imitation of facial and manual gestures by human neonates. *Science, 198*(4312), 74–78.

Meyer, I. H. (2013). Prejudice, social stress, and mental health in lesbian, gay, and bisexual populations: Conceptual issues and research evidence. *Psychology of Sexual Orientation and Gender Diversity, 1*(S), 3–26.

Miller, G. A. (2010). Mistreating psychology in the decades of the brain. *Psychological Science, 5*(6), 716–743.

Moffatt, B. (2011). Responsible authorship: Why researchers must forgo honorary authorship. *Accountability in Research, 18*(2), 76–90.

Montoya, J. G., Holmes, T. H., Anderson, J. N., Maecker, H. T., Rosenberg-Hasson, Y., Valencia, I. J., ... Davis, M. M. (2017). Cytokine signature associated with disease severity in chronic fatigue syndrome patients. *PNAS, 114*(34), E7150–E7158.

National Task Force (1994). *The London Criteria: Report on Chronic Fatigue Syndrome (CFS), Post Viral Fatigue Syndrome (PVFS) and Myalgic Encephalomyelitis (ME)*. Bristol: National Task Force.

Nieuwenhuis, S., Forstmann, B. U., and Wagenmakers, E.-J. (2011). Erroneous analyses of interactions in neuroscience: A problem of significance. *Nature Neuroscience, 14*(9), 1105–1107.

Nosek, B. A., and Bar-Anan, Y. (2012). Scientific utopia: I. Opening scientific communication. *Psychological Inquiry, 23*, 217–243.

Nosek, B. A., Spies, J. R., and Motyl, M. (2012). Scientific utopia: II. Restructuring incentives and practices to promote truth over publishability. *Perspectives on Psychological Science, 7*(6), 615–631.

O'Keeffe, C., and Wiseman, R. (2005). Testing alleged mediumship: Methods and results. *British Journal of Psychology, 96*, 165–179.

Oostenbroek, J., Suddendorf, T., Nielsen, M., Redshaw, J., Kennedy-Costantini, S., Davis, J., ... Slaughter, V. (2016). Comprehensive longitudinal study challenges the existence of neonatal imitation in humans. *Current Biology, 26*(10), 1334–1338.

Open Science Collaboration (2015). Estimating the reproducibility of psychological science. *Science, 349*(6251), aac4716.

Palmer, K. M. (2016, March 3). *Psychology is in crisis over whether it's in crisis*. Retrieved April 1, 2017, from Wired: www.wired.com/2016/03/psychology-crisis-whether-crisis/.

Panter, M. (2017, November 1). *Ghost authorship*. Retrieved November 1, 2017, from American Journal Experts: www.aje.com/en/arc/ghost-authorship/.

Pashler, H., and Harris, C. R. (2012). Is the replicability crisis overblown? Three arguments examined. *Perspectives on Psychological Research, 7*(6), 531–536.

Pelham, B. W., Mirenberg, M. C., and Jones, J. T. (2002). Why Susie sells seashells by the seashore: Implicit egotism and major life decisions. *Journal of Personality and Social Psychology, 82*, 469–487.

Petrie, K. J., and Weinman, J. (2017). The PACE trial: It's time to broaden perceptions and move on. *Journal of Health Psychology, 22*(9), 1198–1200.

Poldrack, R. (2017). The risks of reading the brain. *Nature, 541*, 156.

Poldrack, R. A., Baker, C. I., Durnez, K., Gorgolewski, K. J., Matthews, P. M., Munafò, M. R., ... Yarkoni, T. (2017). Scanning the horizon: Towards transparent and reproducible neuroimaging research. *Nature Reviews Neuroscience, 18*, 115–126.

Poli, R., and Agrimi, E. (2012). Internet addiction disorder: Prevalence in an Italian student population. *Nordic Journal of Psychiatry, 66*(1), 55–59.

Popper, K. (1932). Letter to Egon Friedell, 30 June, 1932 (Engl. transl. by TS). *Hoover Institute Archives: Karl Popper Papers, box 297*, file 22.

Popper, K. R. (1965). *Conjecture and Refutation: The Growth of Scientific Knowledge*. New York, NY: Harper and Row.

Powell, K. (2016). Does it take too long to publish research? *Nature, 530,* 148–151.

Ranehill, E., Dreber, A., Johannesson, M., Leiberg, S., Sul, S., and Weber, R. A. (2015). Assessing the robustness of power posing: No effect on hormones and risk tolerance in a large sample of men and women. *Psychological Science, 26*(5), 653–656.

Reid, A. (2016, January 6). *Scientific study proves scientific studies can't prove anything*. Retrieved from Huffington Post: www.huffingtonpost.com/ann-reid/scientific-study-proves-s_b_10247072.html.

Řezáč, P., Viziová, P., Dobešová, M., Havlíček, Z., and Pospíšilová, D. (2011). Factors affecting dog–dog interactions on walks with their owners. *Applied Animal Behaviour Science, 134,* 170–176.

Risen, J. (2014). *Pay Any Price: Greed, Power, and Endless War*. New York, NY: Mariner Books.

Rosenthal, R. (1979). The 'file drawer problem' and tolerance for null results. *Psychological Bulletin, 86*(3), 638–641.

Sample, I. (2015, August 27). Study delivers bleak verdict on validity of psychology experiment results. Retrieved from Guardian: www.theguardian.com/science/2015/aug/27/study-delivers-bleak-verdict-on-validity-of-psychology-experiment-results.

Sarason, I. G., Sarason, B. R., Shearin, E. N., and Pierce, G. R. (1987). A brief measure of social support: Practical and theoretical implications. *Personal Relationships, 4*(4), 497–510.

Sashidharan, S. P. (2001). Institutional racism in British psychiatry. *Psychiatric Bulletin, 25,* 244–247.

Savitz, J. B., Rauch, S. L., and Drevets, W. C. (2013). Clinical application of brain imaging for the diagnosis of mood disorders. *Molecular Psychiatry, 18,* 528–539.

Schachter, S., and Singer, J. (1962). Cognitive, social, and physiological determinants of emotional state. *Psychological Review, 69,* 379–399.

Schienle, A., Schaefer, A., Stark, R., and Vaitl, D. (2009). Long-term effects of cognitive behavior therapy on brain activation in spider phobia. *Psychiatry Research: Neuroimaging, 172,* 99–102.

Schimmack, U. (2016, April 18). *Replicability report no. 1: Is ego-depletion a replicable effect?* Retrieved April 1, 2017, from Replication Index: https://replicationindex.wordpress.com/2016/04/18/is-replicability-report-ego-depletionreplicability-report-of-165-ego-depletion-articles/.

Schnall, S., Benton, J., and Harvey, S. (2008). With a clean conscience: Cleanliness reduces the severity of moral judgments. *Psychological Science, 19,* 1219–1222.

Schwartz, S. J., Lilienfeld, S. O., Meca, A., and Sauvigné, K. C. (2016). The role of neuroscience within psychology: A call for inclusiveness over exclusiveness. *American Psychologist, 71*(1), 52–70.

Sharpe, M. C., Archard, L. C., Banatvala, J. E., Broysiewicz, L. K., Clare, A. W., David, A., … Lane, R. J. (1991). A report: Chronic fatigue syndrome – Guidelines for research. *Journal of the Royal Society of Medicine, 84*(2), 118–121.

Shaw, G. B. (1921). *Back to Methuselah: A Metabiological Pentateuch.* London: Constable.

Shead, S. (2017, February 27). *The Japanese tech billionaire behind SoftBank thinks the 'singularity' will occur within 30 years.* Retrieved April 1, 2017, from Business Insider UK: http://uk.businessinsider.com/softbank-ceo-masayoshi-son-thinks-singularity-will-occur-within-30-years-2017-2.

Shek, D. T., and Yu, L. (2012). Internet addiction phenomenon in early adolescents in Hong Kong. *Scientific World Journal, 2012,* 104304.

Sherman, S. M., Cheng, Y.-P., Fingerman, K. L., and Schnyer, D. M. (2016). Social support, stress and the aging brain. *Social Cognitive and Affective Neuroscience, 11*(7), 1050–1058.

Shulman, S. (2006). *Undermining Science: Suppression and Distortion in the Bush Administration.* Berkeley, CA: University of California Press.

Simmons, J. P., Nelson, L. D., and Simonsohn, U. (2011). False-positive psychology: Undisclosed flexibility in data collection and analysis allows presenting anything as significant. *Psychological Science, 22*(11), 1359–1366.

Simonsohn, U. (2011). Spurious? Name similarity effects (implicit egotism) in marriage, job, and moving decisions. *Journal of Personality and Social Psychology, 101*(1), 1–24.

Singh, S. P., Greenwood, N., White, S., and Churchill, R. (2007). Ethnicity and the Mental Health Act 1983: Systematic review. *British Journal of Psychiatry, 191,* 99–105.

Skibba, R. (2016). The polling crisis: How to tell what people really think. *Nature, 538,* 304–306.

Smaldino, P. E., and McElreath, R. (2016). The natural selection of bad science. *Royal Society Open Science, 3,* 160384.

Smith, G. (2016). Hurricane names: A bunch of hot air? *Weather and Climate Extremes, 12,* 80–84.

Son, M. (2017, February 28). *Masayoshi Son Keynote at MWC 2017.* Retrieved April 1, 2017, from YouTube: www.youtube.com/watch?v=jIEfPlvnLFw.

Spellman, B. A. (2015). A short (personal) future history of Revolution 2.0. *Perspectives on Psychological Science, 10*(6), 886–899.

Stouten, B. (2017). PACE-GATE: An alternative view on a study with a poor trial protocol. *Journal of Health Psychology, 22*(9), 1192–1197.

Strack, F., Martin, L. L., and Stepper, S. (1988). Inhibiting and facilitating conditions of the human smile: A nonobtrusive test of the facial feedback hypothesis. *Journal of Personality and Social Psychology, 54*, 768–777.

Stroebe, W., and Hewstone, M. (2015, September 17). *What have we learned from the Reproducibility Project?* Retrieved April 1, 2017, from Times Higher Education: www.timeshighereducation.com/opinion/reproducibility-project-what-have-we-learned.

Sturm, T., and Mülberger, A. (2012). Crisis discussions in psychology: New historical and philosophical perspectives. *Studies in History and Philosophy of Biological and Biomedical Sciences, 43*, 425–433.

Thomas, C. R., and Pope, K. (Eds). (2013). *The Origins of Antisocial Behavior*. New York, NY: Oxford University Press.

Timmer, J. (2016, January 7). *Software faults raise questions about the validity of brain studies*. Retrieved April 1, 2017, from Ars Technica: https://arstechnica.com/science/2016/07/algorithms-used-to-study-brain-activity-may-be-exaggerating-results/.

Trafimow, D., and Marks, M. (2015). Editorial. *Basic and Applied Social Psychology, 37*, 1–2.

Vaidya, A. R., and Fellows, L. K. (2015). Testing necessary regional frontal contributions to value assessment and fixation-based updating. *Nature Communications, 6*(10120), 1–12.

van der Linden, S., Leiserowitz, A., Rosenthal, S., and Maibach, E. (2017). Inoculating the public against misinformation about climate change. *Global Challenges, 1*, 1600008.

Viegas, J. (2011, November 3). *Dogs walked by men are more aggressive*. Retrieved November 7, 2011, from Discovery News: http://news.discovery.com/animals/dog-walking-behavior-111103.html.

Villiers de l'Isle-Adam, A. (1886). *L'Ève Future*. Paris: Ancienne Maison Monnier.

Vinge, V. (1993). The coming technological singularity: How to survive in the post-human era. In S. Bailey, R. Ziegfeld, G. A. Landis, J. Chen, L. D. Nichols, and V. Hassett (Eds), *Vision-21: Interdisciplinary Science and Engineering in the Era of Cyberspace* (pp. 11–22). Cleveland, OH: NASA.

Vink, M. (2017). PACE trial authors continue to ignore their own null effect. *Journal of Health Psychology, 22*(9), 1134–1140.

Voutier, C. (2017, January 23). *Beall's list of predatory publishers*. Retrieved April 1, 2017, from Exploring the Evidence Base: https://clinicallibrarian. wordpress.com/2017/01/23/bealls-list-of-predatory-publishers/.

Vul, E., Harris, C., Winkielman, P., and Pashler, H. (2009). Puzzlingly high correlations in fMRI studies of emotion, personality, and social cognition. *Perspectives on Psychological Science, 4*(3), 274–290.

Wagenmakers, E.-J., Beek, T., Dijkhoff, L., Gronau, Q. F., Acosta, A., Adams Jr, R. B., ... Zwaan, R. A. (2016). Registered replication report: Strack, Martin, and Stepper (1988). *Perspectives on Psychological Science, 11*(6), 917–928.

Walters, J. (2016, January 6). *Luminosity fined millions for making false claims about brain health benefits*. Retrieved April 1, 2017, from Guardian: www.theguardian.com/technology/2016/jan/06/lumosity-fined-false-claims-brain-training-online-games-mental-health.

Ware, J. E., and Sherbourne, C. D. (1992). The MOS 36-item short-form health survey (SF-36). *Medical Care, 30*(6), 473–483.

Ware, M., and Mabe, M. (2015). *The STM Report: An Overview of Scientific and Scholarly Journal Publishing* (4th ed.). The Hague: IASTMP.

Wasserstein, R. L., and Lazar, N. A. (2016). The ASA's statement on p-values: Context, process, and purpose. *American Statistician, 70*(2), 129–133.

Watters, E. (2010). *Crazy Like Us: The Globalization of the American Psyche*. New York, NY: Free Press.

Weisberg, D. S., Keil, F. C., Goodstein, J., Rawson, E., and Gray, J. R. (2008). The seductive allure of neuroscience explanations. *Journal of Cognitive Neuroscience, 20*, 470–477.

Weiss, S. L. (2006). Female-specific color is a signal of quality in the striped plateau lizard (*Sceloporus virgatus*). *Behavioral Ecology, 17*(5), 726–732.

White, P. D., Goldsmith, K. A., Johnson, A. L., Potts, L., Walwyn, R., DeCesare, J. C., ... Sharpe, M. (2011). Comparison of adaptive pacing therapy, cognitive behaviour therapy, graded exercise therapy, and specialist medical care for chronic fatigue syndrome (PACE): A randomised trial. *Lancet, 377*, 823–836.

White, P. D., Goldsmith, K., Johnson, A. L., Chalder, T., and Sharpe, M. (2013). Recovery from chronic fatigue syndrome after treatments given in the PACE trial. *Psychological Medicine, 43*, 2227–2235.

Wicherts, J. M., Bakker, M., and Molenaar, D. (2011). Willingness to share research data is related to the strength of the evidence and the quality of reporting of statistical results. *PLoS ONE, 6*(11), e26828.

Williams, L. E., and Bargh, J. A. (2008). Experiencing physical warmth promotes interpersonal warmth. *Science, 322*, 306–307.

Willy, R. (1897). Die Krisis in der Psychologie. *Vierteljahrsschrift für wissenschaftliche Philosophie, 21*, 227–353.

Wilshire, C., Kindlon, T., Matthees, A., and McGrath, S. (2017). Can patients with chronic fatigue syndrome really recover after graded exercise or cognitive behavioural therapy? A critical commentary and preliminary re-analysis of the PACE trial. *Fatigue: Biomedicine, Health and Behavior, 5*(1), 43–56.

Wolpert, L. (1992). *The Unnatural Nature of Science.* London: Faber and Faber.

Xu, X., Aron, A., Brown, L., Cao, G., Feng, T., and Weng, X. (2011). Reward and motivation systems: A brain mapping study of early-stage intense romantic love in Chinese participants. *Human Brain Mapping, 32*, 249–257.

Yarkoni, T. (2016, October 1). *There is no 'tone' problem in psychology.* Retrieved April 1, 2017, from talyarkoni.org: www.talyarkoni.org/blog/2016/10/01/there-is-no-tone-problem-in-psychology/.

Yarritu, I., Matute, H., and Vadillo, M. A. (2014). Illusion of control: The role of personal involvement. *Experimental Psychology, 61*, 38–47.

YouGov. (2012, December 18). *YouGov Omnibus Poll.* Retrieved April 1, 2017, from Huffington Post: http://big.assets.huffingtonpost.com/ghosttoplines.pdf.

Young, K. (1998). *Caught in the Net.* New York: Wiley.

Zhong, C. B., and Liljenquist, K. (2006). Washing away your sins: Threatened morality and physical cleansing. *Science, 46*, 859–862.

Zhu, Y., Zhang, L., Fan, J., and Han, S. (2007). Neural basis of cultural influence on self representation. *Neuroimage, 34*, 1310–1317.

Zöllner, S., and Pritchard, J. K. (2007). Overcoming the winner's curse: Estimating penetrance parameters from case-control data. *American Journal of Human Genetics, 80*, 605–615.

Zung, W. W. (1965). A self-rating depression scale. *Archives of General Psychiatry, 12*, 63–70.

Index

academic psychology departments, 101, 148
academic publication, 153–155, 168–170
    authorship ethics, 160–163
    file-drawer effect, 21
    metrics, 163–165
    peer-review, 142, 155–159, 169–171
    predatory publishers, 159–160
American Statistical Association, 97
American Psychiatric Association, 56
American Psychological Association, 38, 146
    torture controversy, 146
Applied Animal Behaviour Science (journal), 15
artificial intelligence , 46–47
Association for Psychological Science, 31

Basic and Applied Social Psychology (journal),
        97
behaviourism, *see* psychology, schools of
biological psychology, *see* psychology, schools
        of
brain imaging, 11, 23, 53, 121–132
    fMRI (functional magnetic resonance
        imaging), 54, 123–126, 128–129, 130
British Medical Journal, 133
British Psychological Society, 43
Burt, Cyril, 31–32

China, 53, 76, 92
    Chinese language, 108–109, 110–111
    Hong Kong, 52, 117
chronic fatigue syndrome, *see* ME/CFS
CNN, 2
cognitivism, *see* psychology, schools of,
Cohen, Jacob, 96
complementary medicine, 147
confirmation bias, 10, 14, 15, 22, 41
conflation, *see* questionable research practices
cortisol, 16
critical psychology, *see* psychology, specialisms

cross-sectional research, 15
culture differences, 111–113

declinism, 145
depression, *see* mental health
Discovery Channel, 15
divorce, 55
dogs, 15, 16

ego depletion, 17
Einstein, Albert, 6, 48
embodied cognition, 18, 30
English language, 108–110
evolution, 34, 102–105

facial feedback effects, 17
fake news, 28, 29–30
falsification, 7
false-positives, *see* questionable research
        practices
Feynman, Richard, 37–38
file-drawer effect, *see* academic publication
fMRI (functional magnetic resonance imaging),
        see brain imaging
Freud, Sigmund, 36, 107
Fulford, Bill, 114

gender differences, 98–99
Gilbert, Daniel, 24, 25
grounded theory, 41

HARKing (hypothesizing after the results are
        known), *see* questionable research
        practices
health psychology, *see* psychology, specialisms
Heidegger, Martin, 147
Hewstone, Miles, 24
Homo sapiens, 113
Hong Kong, *see* China

How to Lie with Statistics (book), 100
Huffington Post, 4
humanistic psychology, *see* psychology, schools of
hurricanes, 1–3

implicit egotism, 18, 30
International Council of Medical Journal Editors, 161
internet, 52–53
  internet addition, *see* mental health
IQ (intelligence quotient), 47–50, 69
Ireland, 85
Italy, 52

Journal of Experimental Psychology: Learning, Memory, and Cognition, 13, 14
Journal of Personality and Social Psychology, 13, 14, 107

Kuhn, Thomas, 147

Landon, Alfred, 74
Le Monde, 4
Literary Digest (magazine), 74

margin of error, *see* statistics
Maslow, Abraham, 107
ME/CFS (myalgic encephalomyelitis/chronic fatigue syndrome), 132–140
Meehl, Paul, 11
menstruation, 103
mental health
  conditions
    anorexia nervosa, 116–117
    anxiety, 56
    attention deficit hyperactivity disorder, 116
    bulimia, 56–57
    culture-specific conditions, 116
    depression, 34–36, 44, 63, 70, 122–123, 130
    internet addiction, 51–53
    post-traumatic stress disorder, 116
    schizophrenia, 130
  diagnostics, 56–57, 63–64, 114–117
    Beck Depression Inventory, 63
    Differential Emotions Scale, 66

DSM (Diagnostic and Statistical Manual of Mental Disorders), 56–57
    Internet Addiction Test, 52
    Social Readjustment Rating Scale, 54
    Satisfaction With Life Scale, 66
  Mental Health Act (UK), 115
  psychotherapy, 60-61, 132–140, 139–140
    cognitive behaviour therapy (CBT), 134, 136–137, 139
    graded exercise therapy (GET), 134, 136–137, 139
MMR vaccine, 30
myalgic encephalomyelitis, *see* ME/CFS

National Academy of Medicine, 132
National Institute for Health and Care Excellence, 134
Nature (journal), 5, 125
Nature Neuroscience, 125
New York Times, 4
Nosek, Brian, 13, 14

Open Science Collaboration, 5, 6, 13, 24, 25

p-hacking, *see* questionable research practices
PACE Trial, *see* ME/CFS
Philosophy of Science (journal), 11
philosophy, 40
political elections, 73
  electoral opinion polls, 73–80, 82
Popper, Karl, 37, 147
positive emotion, 66
post-truth, 28–29, 31
power posing, 16, 17
priming, 18
Proceedings of the National Academy of Sciences, 1
protanomaly, 120
psychoanalysis, *see* psychology, schools of
psychobabble, 3–4
Psychological Science (journal), 13, 14
psychology
  applications
    psychological measurement, *see* research methods
    psychotherapy, *see* mental health

psychology (*Cont.*)
  liberal bias, 151–152
  paradigms, 31
  reproducibility, *see* replication in psychology
  schools of, 34–36
    behaviourism, 9, 34–35
    biological psychology, 32, 34
    cognitivism, 32, 35
    humanistic psychology, 32, 36
    psychoanalysis, 9, 36, 37–38
    social psychology, 30, 32, 35–36
  specialisms
    critical psychology, 23, 31
    health psychology, 34
    neuroscience, 37, 126–132
    psychophysics, 9
psychophysics, *see* psychology, specialisms
psychotherapy, *see* mental health
p-values, 21, 22, 23, 85–86, 89–91, 93–94, 96, 97, 99

qualitative psychology, *see* research methods
'quantitative' psychology, *see* research methods
questionable research practices, 12
  conflation, 64–67
  false-positives, 17, 19, 61
  HARKing (hypothesizing after the results are known), 10, 13, 42
  p-hacking, 87

Rampant Methodological Flexibility, 10–11, 12, 14, 20, 21, 22, 86, 125, 135–136
randomized controlled trials (RCTs), *see* research methods
reliability, 51, 68–71
  test-retest reliability, 69–70
replication in psychology
  replication crisis, 5, 7–9, 14, 17, 25, 30, 44–45
  contributory factors, 19–24
  crisis denial, 24, 148–150
  early concerns, 8–9, 145–146
  media coverage, 4–5
Reproducibility Project, *see* Open Science Collaboration
research fraud, *see* questionable research practices

research methods
  blinding, 136–137
  convenience samples, 105
  meta-analysis, 24
  mixed methods, 43
  observational research, 15
  psychological measurement, 50–51
  qualitative psychology, 23, 39–44
  'quantitative' psychology, 40
  randomized controlled trials (RCTs), 25
  sample sizes, 23, 76, 94–95, 125
  sampling, 105–107
robots, 46
Roosevelt, Franklin D., 74
Rosenthal, Robert, 21
RUR: Rossum's Universal Robots (film), 46

Science (journal), 5, 125
sexuality, 102, 151
  fertility signalling, 102–105, 120
  marriage, 54–56
  romantic love, 127–128
  virginity, 56
Shaw, George Bernard, 73
singularity (technological), 47
social psychology, *see* psychology, schools of
social support, 65
Son, Masayoshi, 47
statistics, 30, 75–76
  analysis of variance (ANOVA), 84–86
  assumptions of statistical tests, 93
  confidence interval, 76–80
  descriptive statistics, 81
  effect size, 14, 23, 96–97, 138
  inferential statistics, 81
  innumeracy, 76, 87, 100–101, 148
  margin of error, 70, 75–80
  NHST (null-hypothesis significance testing), 88–89, 91–94
  standard deviation, 77, 83–84
  statistical errors, 11–12, 22, 125
  statistical power, 120, 125
  statistical significance, *see* *p*-values
  t-test, 84–86
stress, 32, 53–56, 65–66
  stress response, 32–34
Stroebe, Wolfgang, 24
surveys, 147

TED talk, 16
The Beatles, 87
The Future Eve (novel), 46
The Guardian, 4
The Independent, 4
The Lancet, 30
Times Higher Education, 24

United Kingdom, 75
United States, 55, 74, 98, 108

validity, 51, 53
   construct validity, 57–60
   external validity, 67–68
   internal validity, 60–64

ventromedial prefrontal cortex (VPC),
   128
verification, 7
Villiers de l'Isle-Adam, Auguste, 46
visual perception, 111

Washington Post, 4
'WEIRD' populations, 108,
   113–114
Wikipedia, 34, 40
Wilde, Oscar, 28
winner's curse, 126
Wolpert, Lewis, 37

Yarkoni, Tal, 166–167

Printed in the USA
CPSIA information can be obtained
at www.ICGtesting.com
LVHW010401161223
766494LV00006B/237

9 781352 003000